2

CLAST Manual

James Wooland

with contribution by Liz Dameron

Thinking
Mathematically

4 Edition

Blitzer

ST. PETERSBURG COLLEGE

PEARSON

Prentice
Hall

Upper Saddle River, NJ 07458

Editor-in-Chief: Sally Yagan
Acquisitions Editor: Chuck Synovec
Supplement Editor: Joanne Wendelken
Executive Managing Editor: Kathleen Schiaparelli
Assistant Managing Editor: Karen Bosch Petrov
Production Editor: Robert Merenoff
Supplement Cover Manager: Paul Gourhan
Supplement Cover Designer: Victoria Colotta
Manufacturing Buyer: Ilene Kahn
Manufacturing Manager: Alexis Heydt-Long

© 2008 Pearson Education, Inc.
Pearson Prentice Hall
Pearson Education, Inc.
Upper Saddle River, NJ 07458

The author and publisher of this book have used their best efforts in preparing this book. These efforts include the development, research, and testing of the theories and programs to determine their effectiveness. The author and publisher make no warranty of any kind, expressed or implied, with regard to these programs or the documentation contained in this book. The author and publisher shall not be liable in any event for incidental or consequential damages in connection with, or arising out of, the furnishing, performance, or use of these programs.

Printed in the United States of America

10 9 8 7 6 5 4 3 2 1

ISBN 0-13-175211-1

Pearson Education Ltd., *London*
Pearson Education Australia Pty. Ltd., *Sydney*
Pearson Education Singapore, Pte. Ltd.
Pearson Education North Asia Ltd., *Hong Kong*
Pearson Education Canada, Inc., *Toronto*
Pearson Educación de Mexico, S.A. de C.V.
Pearson Education—Japan, *Tokyo*
Pearson Education Malaysia, Pte. Ltd.

TABLE OF CONTENTS

PART I: A Guide to the CLAST SKILLS

Arithmetic Skills

Geometry Skills

Algebra Skills

Statistics Skills

Logic Skills

Part II: CLAST Exercises for Sections of Thinking Mathematically

Forward

This book is intended to be used in conjunction with Bob Blitzer's liberal arts mathematics text, *Thinking Mathematically*. This is not a stand-alone text.

The content of this book is divided into two parts. The first part is a guide to the specific skills that may be covered on the computational part of the CLAST. The second part is a collection of CLAST-style exercises linked to appropriate sections of *Thinking Mathematically*, along with two 55-question practice exams.

The CLAST Skills

The State of Florida has created a list of 55 skills that may be included on the computations portion of the CLAST. For each of these skills there are specifications that govern such things as how the test problems should be phrased, their level of complexity and style of notation, and the attributes of both the correct and incorrect multiple-choice responses. The reader may refer to the first part of this book for a skill-by-skill description of each of these CLAST items, including worked-out examples. These CLAST skills are labeled and titled in accordance with the State of Florida's definitions.

The Exercises

The second part of this book is a collection of CLAST-style homework exercises linked to various sections of *Thinking Mathematically*. My intention is that the instructor using this text will assign these CLAST-style homework problems, when appropriate, along with exercises from *Thinking Mathematically*. For exposition of the mathematical concepts involved in these homework problems the student can refer to the indicated section of Bob Blitzer's book, and may also refer to the first part of this book for specific facts about the CLAST skills.

The multiple-choice exercises have been written so that they conform to the state's CLAST item specifications. In cases where the phrasing of a problem seems awkward or nongrammatical, it is probably due to the fact the imperfect phrasing and grammar have been inherited from the state's CLAST specifications. Likewise, there are some instances (see CLAST skills IV.D.1 and IV.D.2, for example) in which certain CLAST problems may be stated in such a way that either the statement of the problem or the intended solution is mathematically incorrect. Again, these errors are not due to carelessness on the part of the author, but rather they result from the author's attempt to write exercises that conform to the state's item specifications, even though those specifications are sometimes flawed.

The Practice Tests

At the end of this book are two practice tests. These tests contain 55 questions each (one question for each of the 55 CLAST skills). The student who is on the verge of taking the CLAST can prepare by taking the first of these two practice tests. The key for this practice test includes not only the correct answers for all of the questions, but also indicates the CLAST skill covered by each question. By referring to the appropriate sections of this book and *Thinking Mathematically*, the student can brush up on the skills covered by problems that he or she answered incorrectly. After further study, the student can then attempt the second practice test. A student should hope to get a score of at least 70% (39 out of 55 questions answered correctly) on the practice tests in order to be in a good position to pass the real exam. On the actual exam, however, not every possible CLAST skill will be covered. A student may take the paper and pencil form of the test which has 55 multiple choice items. If taken on the computer, the math subtest of CLAST consists of roughly 33 questions, again multiple choice.

Acknowledgements

I'd like to thank Bob Blitzer for writing his wonderful book, Sally Yagan at Prentice Hall for making this project possible and enduring my complaints, and Audra Walsh and the Prentice Hall production people for assembling the final product. I'd also like to thank my loving wife Donna for her support as well as for her proofreading skills.

PART I
A GUIDE TO THE CLAST SKILLS

CLAST SKILL I.A.1a
The student will add and subtract rational numbers.

Refer to Section 5.3 of *Thinking Mathematically*.

You will be given two rational numbers and asked to add or subtract. At least one of the two numbers will be a mixed number or proper fraction.

SKILL I.A.1a EXAMPLE

$$6\frac{3}{4} - \frac{2}{3} =$$

A. 6　　　　　　B. $6\frac{1}{12}$　　　　　　C. $6\frac{1}{7}$　　　　　　D. $6\frac{1}{2}$

SOLUTION

Recall that $6\frac{3}{4}$ means $6 + \frac{3}{4}$, so

$$6 + \frac{3}{4} - \frac{2}{3}$$ 　　　　　　*We will combine the two fractions.*

　　　　　　　　　　　　　　　　Their common denominator is 12.

$$= 6 + \frac{3}{4}\cdot\frac{3}{3} - \frac{2}{3}\cdot\frac{4}{4}$$

$$= 6 + \frac{9}{12} - \frac{8}{12}$$

$$= 6 + \frac{1}{12}$$ 　　　　　　*We can write $6 + \frac{1}{12}$ as a mixed number.*

$$= 6\frac{1}{12}$$ 　　　　　　The correct choice is B.

For more practice with this skill go to page 196 of this manual.

CLAST SKILL I.A.1b
The student will multiply and divide rational numbers.

You will be given two rational numbers and asked to multiply or divide. At least one of the two numbers will be a mixed number or proper fraction.

SKILL I.A.1b EXAMPLE

$$10\frac{5}{8} \div 4 =$$

A. $2\frac{5}{32}$ B. $40\frac{1}{2}$ C. $\frac{32}{85}$ D. $2\frac{21}{32}$

SOLUTION

Recall that $10\frac{5}{8}$ means $10 + \frac{5}{8}$. We will write $10\frac{5}{8}$ as an improper fraction before we divide.

$10\frac{5}{8} = 10 + \frac{5}{8}$ *First we will write 10 as a fraction whose denominator is 8.*

$= \frac{80}{8} + \frac{5}{8} = \frac{85}{8}$ *We now return to the original problem.*

$10\frac{5}{8} \div 4 = \frac{85}{8} \div 4$ *Write 4 as a fraction.*

$= \frac{85}{8} \div \frac{4}{1}$ *Invert the divisor and multiply.*

$= \frac{85}{8} \times \frac{1}{4} = \frac{85 \times 1}{8 \times 4} = \frac{85}{32}$ *Write this as a mixed fraction. The greatest multiple of 32 that is less than 85 is 64.*

$= \frac{64}{32} + \frac{21}{32} = 2 + \frac{21}{32}$

$= 2\frac{21}{32}$ *The correct choice is B.*

For more practice with this skill go to page 196 of this manual.

CLAST SKILL I.A.2a
The student will add and subtract rational numbers in decimal form.

Refer to Section 5.3 of *Thinking Mathematically.*

You will be given two decimal numbers and asked to add or subtract. Of course, on the CLAST you will not have a calculator, so you will need to remember that old elementary school method of vertically aligning the two numbers and adding or subtracting.

The two numbers will have values that may range from −20 to 20, and will not necessarily have the same number of decimal places.

In many cases, you may be able to choose the correct answer by estimation.

EXAMPLE A

$4.446 - 2.7 =$

A. 2.746   B. 1.746   C. 2.439   D. 4.419

SOLUTION

$$\begin{array}{r} 4.446 \\ -\ 2.700 \\ \hline 1.746 \end{array}$$

The correct choice is B.

Notice that you could choose the correct answer by estimation:

4.446 is about equal to 4.5, and 2.7 is about equal to 3.

$4.5 - 3 = 1.5$

Choice B is the only answer that is close to 1.5.

For more practice with this skill go to page 197 of this manual.

CLAST SKILL I.A.2a
The student will multiply and divide rational numbers in decimal form.

Refer to Section 5.3 of *Thinking Mathematically*.

You will be given two decimal numbers and asked to multiply or divide. Of course, on the CLAST you will not have a calculator, so you will need to remember methods of multiplication and long division that you learned in elementary school.

The two numbers will have values that may range from −20 to 20, and will not necessarily have the same number of decimal places.

In many cases, you may be able to choose the correct answer by estimation.

EXAMPLE B

$(-.03) \times (-1.5) =$

A. −.0045   B. .0045   C. −.045   D. .045

Since we are multiplying two negative numbers the product will be positive. We can obtain the correct product by multiplying the absolute values of these two numbers. The two numbers have a total of 3 digits to the right of the decimal point, we will multiply two numbers without decimal points, obtain the product, and insert the decimal point so that there are three digits to the right of the decimal point.

EXAMPLE B SOLUTION

$$\begin{array}{r} 15 \\ \times\ \ 3 \\ \hline 45 \end{array}$$

We move the decimal point three places to the left: .045

The answer is positive.

The correct choice is D.

For more practice with this skill go to page 197 of this manual.

CLAST SKILL I.A.3
The student will demonstrate the ability to calculate percent increase and percent decrease.

Refer to Section 8.1 of *Thinking Mathematically*.

This skill may be tested in one of two possible formats. In one case you may be given a description of a quantity that is increasing or decreasing from one specified value to another specified value, and asked to compute either the percent increase or percent decrease, as appropriate. In the other case, you will be given a specified value and asked to either increase or decrease that value by a specified percent.

In these problems the specified values will be positive integers no greater than 1000, and the percents will be positive integers less than 200%.

EXAMPLE A

If 750 is decreased to 600, what is the percent decrease?

A. 125%   B. 150%   C. 20%   D. 2%

EXAMPLE A SOLUTION

For this decreasing quantity, the original amount is 750 and the ending amount is 600.

As a *decimal number*,

$$\text{percent decrease} = \frac{\text{amount of decrease}}{\text{original amount}} = \frac{\text{original amount - ending amount}}{\text{original amount}}$$

$$= \frac{750 - 600}{750} = \frac{150}{750}$$

$$= .2$$

Now convert to a percent:

$.2 = 20\%$

The correct choice is C.

SKILL I.A.3

EXAMPLE B

If 64 is increased to 112, find the percent increase.

A. 48%   B. 43%   C. 57%   D. 75%

EXAMPLE B SOLUTION

For this increasing quantity, the original amount is 64 and the ending amount is 112.

As a *decimal number*,

$$\text{percent increase} = \frac{\text{amount of increase}}{\text{original amount}} = \frac{\text{ending amount - original amount}}{\text{original amount}}$$

$$= \frac{112 - 64}{64} = \frac{48}{64} = \frac{48 \div 16}{64 \div 16} = \frac{3}{4}$$

$$= .75$$

Now convert to a percent: $.75 = 75\%$.   The correct choice is D.

EXAMPLE C

If you increase 72 by 25% of itself, what is the result?

A. 18   B. 90   C. 97   D. 288

EXAMPLE C SOLUTION

First we will find 25% of 72, using the fact that 25% = .25 as a decimal number.

25% of 72 = (.25)(72) = 18

Now, if 72 is increased by 25% of itself, it is increased by 18.

72 + 18 = 90

The correct choice is B.

For more practice with this skill go to page 236 of this manual.

CLAST SKILL I.A.4
The student will solve the sentence a% of b is c, where values for two of the variables are given.

Refer to Section 8.1 of *Thinking Mathematically*.

As the title of the skill indicates, you will be given a statement which involves some variation of the statement "*a*% of *b* is *c*," along with values for two of the three variables. You will be asked to solve for the remaining variable. This skill may test your ability to convert between decimal numbers, fractions and percents, as well as to perform rudimentary algebraic manipulations.

In these problems a percent may be a whole number, like 45%, or a decimal number with no more than one digit to the right of the decimal point, like 64.2%, or a mixed number like $10\frac{3}{4}\%$.

All percent values will be less than 200%.

Other numbers will be positive rational numbers less than 500 and may contain no more than two digits to the right of the decimal point.

EXAMPLE A

What is 18% of 60?

A. 33.3   B. 10.8   C. 42   D. 333

EXAMPLE A SOLUTION

To find 18% of 60, we write 18% in its equivalent decimal form and multiply by 60.

Recall that to convert a percent to a decimal number, we move the decimal point two places to the left and drop the percent sign.

As a decimal number, 18% = .18.

Thus, 18% of 60 = (.18)(60) = 10.8.

The correct choice is B.

SKILL I.A.4

EXAMPLE B

48 is what percent of 240?

A. 115.2%   B. 5%   C. .2%   D. 20%

EXAMPLE B SOLUTION

To answer this question we divide 48 by 240 and express the result as a decimal number, and then convert that decimal number to an equivalent percent.

$$\frac{48}{240} = .2 = 20\%$$

The correct choice is D.

EXAMPLE C

64 is 40% of what number?

A. 25.6   B. 16   C. 160   D. 256

EXAMPLE C SOLUTION

Let x be the number.

Recall that as a decimal number, $40\% = .4$.

Then

$.4x = 64$

Solving for x, we have

$$x = \frac{64}{.4} = 160$$

The correct choice is D.

For more practice with this skill go to page 235 of this manual.

CLAST SKILL II.A.1
The student will recognize the meaning of exponents.

Refer to Section 5.2 of *Thinking Mathematically.*

You will be given either a single positive rational number in exponential form, or an expression involving the addition, subtraction, multiplication or division of two positive rational numbers (one or both of which may be in exponential form).

The exponents will be integers from 1 to 5.

The problems will not involve computing "answers." Rather, they will test your basic understanding of the meaning of exponents, by asking you to select another expression that is equivalent to the given expression.

EXAMPLE A

$8^5 \div 3^4 =$

A. $(5)^1$

B. $(5 \cdot 5 \cdot 5 \cdot 5 \cdot 5 \cdot 5 \cdot 5 \cdot 5) \div (4 \cdot 4 \cdot 4)$

C. $(8 \cdot 8 \cdot 8 \cdot 8 \cdot 8) \div (3 \cdot 3 \cdot 3 \cdot 3)$

D. $\left(\dfrac{8}{3}\right)^{5/4}$

EXAMPLE A SOLUTION

In this exercise you aren't asked to find the value of $8^5 \div 3^4$ or to simplify that expression. You are merely asked to choose another expression that means the same thing as $8^5 \div 3^4$, based on your understanding of exponents.

The meaning of 8^5 is "five factors of 8 multiplied together." $8^5 = (8)(8)(8)(8)(8)$.

The meaning of 3^4 is "four factors of 3 multiplied together." $3^4 = (3)(3)(3)(3)$.

So, the meaning of $8^5 \div 3^4$ is $(8 \cdot 8 \cdot 8 \cdot 8 \cdot 8) \div (3 \cdot 3 \cdot 3 \cdot 3)$.

The correct choice is C.

SKILL II.A.1

EXAMPLE B

$$\left(\frac{3}{4}\right)^3 =$$

A. $\left(\dfrac{3 \times 3 \times 3}{4}\right)$

B. $3\left(\dfrac{3}{4}\right)$

C. $\dfrac{3}{4} + \dfrac{3}{4} + \dfrac{3}{4}$

D. $\dfrac{3}{4} \times \dfrac{3}{4} \times \dfrac{3}{4}$

EXAMPLE B SOLUTION

The meaning of $\left(\dfrac{3}{4}\right)^3$ is "three factors of $\dfrac{3}{4}$ multiplied together."

$\left(\dfrac{3}{4}\right)^3$ means $\dfrac{3}{4} \times \dfrac{3}{4} \times \dfrac{3}{4}$.

Notice again that you weren't asked to evaluate or simplify the given expression. The exercise is more basic than that.

The correct choice is D.

For more practice with this skill go to page 195 of this manual.

CLAST SKILL II.A.2
The student will recognize the role of the base number in determining place value in the base-ten numeration system.

Refer to Section 4.1 of *Thinking Mathematically*.

The problem will have one of these two formats:

1. You will be given a standard base-ten numeral (a "decimal number') and asked to identify the *place value* of a specified digit or select the correct *expanded form* of the numeral;

2. You will be given the *expanded form* of a base-ten numeral and asked to select the correct standard numeral.

EXAMPLE A
Select the place value associated with the underlined digit. 3,95<u>5</u>,069,300.007 455 2

A. 10^7 B. 10^5 C. $\dfrac{1}{10^5}$ D. 10^6

EXAMPLE A SOLUTION
Recall the meaning of place value for a decimal number.

Each digit in a decimal number occupies a place whose value is a power of 10.

The first position to the left of the decimal point is the *ones*, or 10^0 place.
The second position to the left of the decimal point is the *tens*, or 10^1 place.
The third position to the left of the decimal point is the *hundreds*, or 10^2 place.
Generally, the nth position to the left of the decimal point is the 10^{n-1} place.

The first position to the right of the decimal point is the *tenths*, or $\dfrac{1}{10}$ place.

The second position to the right of the decimal point is the hundredths, or $\dfrac{1}{10^2}$ place.

The third position to the right of the decimal point is the thousandths, or $\dfrac{1}{10^3}$ place.

Generally, the nth position to the right of the decimal point is the $\dfrac{1}{10^n}$ place.

The underlined digit in 3,95<u>5</u>,069,300.007 455 2 is in the seventh position to the left of the decimal point, so its place value is 10^6.

The correct choice is D.

11

SKILL II.A.2

EXAMPLE B

Select the correct expanded notation for 600,000.000 8.

A. $\left(6 \times 10^5\right) + \left(8 \times \dfrac{1}{10^4}\right)$

B. $\left(6 \times 10^6\right) + \left(8 \times \dfrac{1}{10^4}\right)$

C. $\left(6 \times 10^6\right) + \left(8 \times \dfrac{1}{10^3}\right)$

D. $\left(6 \times 10^5\right) + \left(8 \times \dfrac{1}{10^3}\right)$

EXAMPLE B SOLUTION

The numeral has a 6 in the sixth place to the left of the decimal point.

This a six in the 10^5 place, and equates to 6×10^5.

The numeral also has an 8 in the fourth place to the right of the decimal point.
This is an 8 in the $\dfrac{1}{10^4}$ place, and equates to $8 \times \dfrac{1}{10^4}$.

Thus, the numeral is equal to $\left(6 \times 10^5\right) + \left(8 \times \dfrac{1}{10^4}\right)$.

The correct choice is A.

EXAMPLE C

Select the numeral for $\left(9 \times 10^3\right) + \left(4 \times 10^1\right) + \left(8 \times \dfrac{1}{10^2}\right)$

A. 904.008

B. 904.08

C. 9,040.08

D. 9,004.08

EXAMPLE C SOLUTION

The numeral must have a 9 in the fourth position to the left of the decimal point, a 4 in the second position to the left of the decimal point, an 8 in the second position to the right of the decimal point, and zeros in the other positions.

This numeral is 9,040.08.

The correct choice is C.

For more practice with this skill go to page 191 of this manual.

CLAST SKILL II.A.3
The student will identify equivalent forms of positive rational numbers involving decimals, percents, and fractions.

Refer to Section 8.1 of *Thinking Mathematically.*

You will be given a positive rational number that may be expressed as a fraction, decimal number or percent. You will be asked to choose an equivalent number. The multiple-choice options may be fractions, decimal numbers or percents.

The decimal numbers in these problems may range from 0.00001 to 200, the numerators and denominators of fractions may range from 1 to 10,000, and percents may range from 0.01% to 10,000%.

The items will include familiar values such as 1/3, 1/2, 25%, 10%, 0.05, 0.25 and so on.

Recall the following facts.

1. *To convert a decimal number to a percent, move the decimal point two places to the right and affix a percent sign.*

$$.15 = 15\% \qquad\qquad .003 = 0.3\% \qquad\qquad 6.8 = 680\%$$

2. *To convert a percent to a decimal number, move the decimal point two places to the left and drop the percent sign.*

$$86.8\% = .868 \qquad\qquad .00072\% = .0000072 \qquad\qquad 195\% = 1.95$$

3. *To convert a fraction to a decimal number, divide the denominator into the numerator, or write the fraction as an equivalent fraction whose denominator is a power of 10 and bear in mind the meaning of tenths, hundredths, thousandths, and so on.*

$$\frac{3}{4} = \frac{3 \times 25}{4 \times 25} = \frac{75}{100} = .75$$

$$\frac{7}{8} = \frac{7 \times 125}{8 \times 125} = \frac{875}{1000} = .875 \text{ or } \frac{7}{8} = 7 \div 8 = .875 \text{ using long division.}$$

$$\frac{23}{50} = \frac{23 \times 2}{50 \times 2} = \frac{46}{100} = .46 \text{ or } \frac{23}{50} = 23 \div 50 = .46 \text{ using long division.}$$

SKILL II.A.3

4. *To convert a decimal number to a fraction, use the meaning of tenths, hundredths, thousandths and so forth to get a fraction whose denominator is power of 10, then reduce the fraction if possible.*

$$0.8 = \frac{8}{10} = \frac{4}{5} \qquad\qquad\qquad\qquad 0.39 = \frac{39}{100}$$

$$1.025 = 1 + 0.025 = 1 + \frac{25}{1000} = 1 + \frac{1}{40} = 1\frac{1}{40}$$

5. *To convert a percent to a fraction, first convert to a decimal number, then convert the decimal number to a fraction.*

$$30\% = .30 = .3 = \frac{3}{10} \qquad\qquad 180\% = 1.80 = 1.8 = 1 + .8 = 1 + \frac{8}{10} = 1 + \frac{4}{5} = 1\frac{4}{5}$$

6. *To convert a fraction to a percent, first convert the fraction to a decimal number, then convert the decimal number to a percent.*

$$\frac{3}{5} = \frac{3 \times 2}{5 \times 2} = \frac{6}{10} = .6 = 60\% \qquad\qquad \frac{9}{40} = \frac{9 \times 25}{40 \times 25} = \frac{225}{1000} = .225 = 22.5\%$$

EXAMPLE A

$4.5 =$

A. .045%   B. $\frac{45}{100}$ C. 4.5%   D. 450%

EXAMPLE A SOLUTION

To convert the number 4.5 to a percent, we move the decimal point two places to the right and affix the percent sign.

$4.5 = 450\%$

The correct choice is D.

EXAMPLE B

0.28% =   A. $\dfrac{28}{100}$ B. 28   C. 0.028   D. $\dfrac{28}{10000}$

EXAMPLE B SOLUTION

To convert this percent to a decimal number, we move the decimal point two places to the left and drop the percent sign.

0.28% = 0.0028

This shows us that choices B and C are incorrect.

To convert the decimal number 0.0028 to a fraction, we recall that this decimal number is read as "28 ten-thousandths." This means that the fractional equivalent is $\dfrac{28}{10000}$.

The correct choice is D.

EXAMPLE C

$1\dfrac{1}{20}$ =   A. 1.2   B. 1.05   C. 120%   D. .00105%

EXAMPLE C SOLUTION

First, recognize that $1\dfrac{1}{20}$ means $1 + \dfrac{1}{20}$.

Now, convert $\dfrac{1}{20}$ to a decimal number. One way to do so is to rewrite $\dfrac{1}{20}$ as an equivalent fraction whose denominator is a power of 10.

$$\frac{1}{20} = \frac{1 \times 5}{20 \times 5} = \frac{5}{100} = .05$$

Thus, $1 + \dfrac{1}{20} = 1 + .05 = 1.05$

The correct choice is B.

If we wanted to write this number as a percent, we would move the decimal point two places to the right and affix a percent sign.

1.05 = 105%. This shows that choices C and D are incorrect.

For more practice with this skill go to page 234 of this manual.

CLAST SKILL II.A.4
The student will determine the order relation between real numbers.

Refer to Sections 5.3 and 5.4 of *Thinking Mathematically*.

You will be given two numbers and asked to place between the two numbers the symbol of equality or inequality ("=" or "<" or ">") that correctly describes the relationship between the two numbers.

The numbers may be mixed numbers, fractions, repeating or terminating decimal numbers, or square roots of positive integers that are not perfect squares.

Remember these basic ideas:

1. Negative numbers are always less than positive numbers.

2. When comparing two negative numbers, compare their absolute values. The lesser negative number will be the one with the greater absolute value.

3. When comparing two positive decimal numbers, identify the leftmost decimal place in which the two numbers differ. The number having the greater digit in that decimal place will be the greater of the two numbers.

4. When comparing two positive fractions, you can use this general fact:
$\dfrac{a}{b} < \dfrac{c}{d}$ if and only if $ad < cb$.

5. When making a comparison involving the square roots of positive numbers, you can use this general fact:

$\sqrt{a} < \sqrt{b}$ if and only if $a < b$.

EXAMPLE A

Identify the symbol that should be placed in the blank to form a true statement.

2.505 _____ 2.5$\overline{05}$ A. =   B. <   C. >

EXAMPLE A SOLUTION

Recall that $2.505 = 2.5050000\overline{0}$ and that $2.5\overline{05} = 2.50505050\overline{5}$.

Comparing $2.50500000\overline{0}$ with $2.505050\overline{50}5$, we see that the leftmost decimal place in which they differ is the fifth position to the right of the decimal point. In that position, the number with the greater digit is $2.505050\overline{50}5$, so 2.505050505 is the greater number.

The correct choice is B.

EXAMPLE B

Identify the symbol that should be placed in the blank to form a true statement.

$\sqrt{24}$ _____ 5.0211   A. =   B. <   C. >

EXAMPLE B SOLUTION

We will make this comparison by comparing the two numbers to a third, more familiar number.

We know that $\sqrt{24} < \sqrt{25}$, since $24 < 25$.

We also know that $\sqrt{25} = 5$.

Finally, we know that $5 < 5.0211$.

Putting these three facts together, we have this result:

$\sqrt{24} < \sqrt{25} < 5.0211$, so $\sqrt{24} < 5.0211$.

The correct choice is B.

EXAMPLE C

Identify the symbol that should be placed in the blank to form a true statement.

$\dfrac{3}{5}$ _____ $\dfrac{7}{12}$ A. =   B. <   C. >

EXAMPLE C SOLUTION

We use the fact that for positive fractions $\dfrac{a}{b} < \dfrac{c}{d}$ if and only if $ad < cb$.

(3)(12) _____ (7)(5)

36 _____ 35   36 > 35   The correct choice is C.

For more practice with this skill go to pages 198 and 202 of this manual.

CLAST SKILL II.A.5
The student will identify a reasonable estimate of a sum, average or product of numbers.

Refer to Section 1.2 of *Thinking Mathematically.*

You will be given a word problem involving a range of data described using either exact or approximate values. The description will specify the smallest value, the largest value, and the number of data points in the distribution.

You will be asked to select a reasonable estimate of either the sum, product or average of the data in the distribution. The correct choice will be the only reasonable answer among the four given choices; all of the other choices will be "clearly not possible" estimates of the sum, product or average.

EXAMPLE A

Franklin has caught two bluefish, five speckled trout and one redfish. The bluefish both weighed approximately 2 pounds, the trout ranged in weight from 2.5 pounds to 4.5 pounds, and the redfish weighed 7 pounds. Which of the following is a reasonable estimate of the total weight of the eight fish?

A. 16 pounds

B. 50 pounds

C. 23.5 pounds

D. 30 pounds

EXAMPLE A SOLUTION

A reasonable estimate might be the following:
$2(2) + 2.5 + 3(3.5) + 4.5 + 7 = 28.5$ pounds.

This estimate is based on the following: 2 bluefish at two pounds each, *plus*
the smallest trout at 2.5 pounds, *plus*
3 other trout at roughly 3.5 pounds each, *plus*
the largest trout at 4.5 pounds, *plus*
the redfish at 7 pounds.

There are other ways to arrive at reasonable estimate, but notice that choice A (16 pounds) is not possible, because that estimate assumes that all of the fish weigh only 2 pounds each.

Choice B (50 pounds) is not reasonable, because all of the trout would have to weigh more than 6 pounds each in order to combine with the other given weights in order to give this total.

Likewise, choice C (23.5 pounds) is not possible, because all of the trout would have to weigh only 2.5 pounds each in order to combine with the other given weights in order to produce this total.

Choice D is the only reasonable estimate.

EXAMPLE B

Three hundred students took a math quiz. All of the students scored less than 20 but more than 6 on the quiz. Which of the following could be a reasonable estimate of the average quiz score?

A. 4

B. 6

C. 12

D. 20

EXAMPLE B SOLUTION

In any collection of numbers (such as quiz scores), as long as the numbers are not all identical, the average must be larger than the smallest number and smaller than the largest number. Since all of these quiz scores are greater than 6 and less than 20, the average must be greater than 6 and less than 20; this eliminates all of the choices except C.

The only reasonable estimate is C.

SKILL II.A.5

EXAMPLE C

Forty-eight men and 12 women auditioned for roles as wrestlers in the Wide World of Wrestling Alliance. Among the men, the smallest weighed 250 pounds and the largest weighed 400 pounds. Among the women, the smallest weighed 135 pounds and the largest weighed 210 pounds. Which of the following is a reasonable estimate of the total combined weight of the men?

A. 15,000 pounds

B. 10,000 pounds

C. 1,000 pounds

D. 20,000 pounds

EXAMPLE C SOLUTION

For estimation purposes, we can round from 48 to 50 wrestlers, and assume an average weight of 300 pounds. $(50)(300) = 15,000$.

15,000 pounds is a reasonable estimate, but there are many other reasonable estimates. Choice B (10,000 pounds) is not reasonable, because if there were only 40 wrestlers and all 40 of them were of the minimum weight (250 pounds), then the total weight would be $(40)(250) = 10,000$ pounds. For this reason, choice B is too small to be reasonable.

Since 10,000 pounds is too small to be reasonable, choice C (1,000 pounds) is way too small to be reasonable.

Choice D is too big to be reasonable, because if there were 50 wrestlers and all of them weighed the maximum of 400 pounds each, then their total weight would be $(50)(400) = 20,000$ pounds.

Choice A is the only reasonable estimate.

For more practice with this skill go to page 167 of this manual.

CLAST SKILL III.A.1
The student infers relations between numbers in general by examining particular number pairs.

Refer to Sections 1.1 and 5.7 of *Thinking Mathematically*.

This skill is primarily a test of your ability to recognize patterns, although some of the problems may involve topics such as arithmetic or geometric sequences. There are two distinct formats for this problem.

In the first format, you will be given a list of ordered pairs of numbers. There will be a specific linear or quadratic relationship between the two numbers in each ordered pair. One of the numbers will be missing from the last pair, and you will be asked to identify that missing number.

The ordered pairs will have one of these forms:

(x, nx) or $(x, x/n)$ where n = 2, 3, 4, or 5;

$(x, x + n)$ or $(x + n, x)$ where $1 \le n \le 10$;

(x^2, x) or (x, x^2).

x will be a rational number.

In the second format, you will be given a sequence of numbers. The type of sequence will be identified as either arithmetic, geometric or harmonic. You will be asked to identify the next number in the sequence.

An **arithmetic** progression is a sequence of numbers in which a fixed value (the **common difference**) is added to each term in order to generate the next term. You can find the common difference by subtracting the first term from the second term.

A **geometric** progression is a sequence of numbers in which a fixed value (the **common ratio**) is multiplied onto each term in order to generate the next term. You can find the common ratio by dividing the first term **into** the second term.

A **harmonic** progression is a sequence of fractions whose **reciprocals** form an **arithmetic** progression.

SKILL III.A.1

EXAMPLE A

Look for a common linear relationship between the numbers in each pair. Then identify the missing term.

(6, 2) (2.4, .8) (–12, –4) (30, 10), (2/3, ___)

A. –10/3

B. 2

C. 3

D. 2/9

EXAMPLE A SOLUTION

We must find a linear relationship that works for all four of the given ordered pairs, and then apply that relationship to the given element of the last ordered pair.

In examining the first ordered pair, (6, 2), one obvious relationship between the two numbers is this: "To produce the second number, we subtract 4 from the first number." However, if we apply this relationship to the next ordered pair, it fails, because 2.4 – 4 is not .8.

Another possible relationship between the two elements of the pair (6, 2) is this: "To produce the second number, we divide the first number by 3."
If we apply this rule to the second ordered pair, it works, because 2.4/3 = .8.
If we apply the rule to the next ordered pair, it still works, because –12/3 = –4.
If we apply this rule to the next ordered pair it still works, because 30/3 = 10.

We find the missing element of the last pair by applying the rule to 2/3:

$$2/3 \div 3 = \frac{2}{3} \times \frac{1}{3} = \frac{2}{9}$$

The correct choice is D.

EXAMPLE B

Identify the missing term from the following harmonic progression.

1/24, 1/20, 1/16, 1/12, 1/8, ____

A. 1/4

B. 1/2

C. 2/3

D. −1/4

EXAMPLE B SOLUTION

A **harmonic** progression is a sequence of numbers whose **reciprocals** form an arithmetic progression. We can find the missing term in the harmonic progression by examining the sequence of reciprocals.

Here is the sequence of reciprocals: 24, 20, 16, 12, 8, ___

Since this is an arithmetic sequence, we can find its common difference by subtracting the first term from the second term:

$20 - 24 = -4$

The common difference is −4. We find the missing term of the arithmetic progression by adding −4 to the last term.

$8 + (-4) = 8 - 4 = 4.$

Since the missing term of the arithmetic progression is 4, the missing term of the harmonic progression is the reciprocal of 4, which is 1/4.

The correct choice is A.

For more practice with this skill go to pages 166 and 204 of this manual.

CLAST SKILL IV.A.1
The student solves real-world problems which do not require the use of variables and do not involve percent.

Refer to Section 5.3 of *Thinking Mathematically.*

You will be given a word problem that may refer to a situation from business, social studies, industry, education, economics, environmental studies, the arts, physical science, sports or a consumer-related context. Solving the problem should not require special knowledge of the field from which the problem is drawn. The problem may contain irrelevant information.

The fact that these problems "do not require the use of variables" suggests that they should be slightly more simple or straightforward than other word problems.

EXAMPLE

Allan has been told that if his weekly dietary intake is 20,000 calories, then he will maintain his present weight of 190 pounds, and that if he reduces his dietary intake by 5000 calories per week, his weight will drop by one pound per week. He estimates that he drinks 10 cans of beer per week, and that each can of beer contains 200 calories. He also drinks 12 cans of soda per week, and each can of soda provides 300 calories. Over the course of six months (26 weeks), how much weight can he expect to lose if he eliminates beer from his diet and makes no other dietary changes?

A. 12.6 pounds   B. 10.4 pounds   C. 65 pounds   D. 179.6 pounds

SOLUTION

Each week Allan's caloric intake from beer is (10 cans)(200 calories per can) = 2000 calories.

Since in his case elimination of 5000 calories per week is supposed to result in one pound of lost weight, his weekly weight loss associated with the elimination of beer from his diet is (2000 calories)/(5000 calories per pound) = .4 pounds.

Over the course of 26 weeks, the total expected weight loss would be
(26 weeks)(.4 pounds per week) = 10.4 pounds.   The correct choice is B.

Notice that we did not use variables to solve this problem; the solution was obtained through a sequence of observations and corresponding arithmetic computations. Also notice that the information about the caloric intake from soda was irrelevant, and likewise the fact that he weighs 190 pounds was not used in this calculation.

For more practice with this skill go to page 199 of this manual.

CLAST SKILL IV.A.2
The student solves real-world problems which do not require the use of variables but do require the use of percent.

Refer to Section 8.1 of *Thinking Mathematically*.

You will be given a real-world word problem whose context involves situations from business, economics, social studies, industry, education, environmental studies, the arts, physical science, sports, or a consumer-related context. Solving the problem will not require special knowledge of the subject area.

The solution will require anywhere from two to seven steps involving addition, subtraction, multiplication or division, and may involve conversions such as percent to decimal or unit conversions. These problems may also include irrelevant information. At least one step will require calculation of a percent.

EXAMPLE A

Warren purchased four pounds of apples at $1.20 per pound, a can of beans for $0.60, a loaf of bread for $1.50, a can of paint for $3.00 and a paintbrush for $2.00. He had to pay 7% sales tax on the non-food items. How much sales tax did he pay?

A. $12.25

B. $0.35

C. $5.35

D. $12.73

EXAMPLE A SOLUTION

The non-food items are the paint and the paintbrush. He spent a total of $5.00 on these two items. The sales tax is then

7% of $5.00

$= (.07)(\$5.00)$

$= \$0.35$

The correct choice is B.

SKILL IV.A.2

EXAMPLE B

After two months of dieting and exercise Jeff's weight dropped from 180 pounds to 171 pounds. What was the percent decrease in his weight?

A. 5%

B. .05%

C. 5.3%

D. 9%

EXAMPLE B SOLUTION

Recall the method used to calculate percent decrease.

As a decimal number,

$$\text{percent decrease} = \frac{\text{amount of decrease}}{\text{original amount}}.$$

We convert this decimal number to a percent by moving the decimal point two places to the right and affixing a percent sign.

In this case, the original weight was 180 pounds and the amount of decrease was 9 pounds (since his weight decreased from 180 to 171 pounds).

$$\text{Percent decrease} = \frac{9}{180} = .05$$

Converting to a percent,

$.05 = 5\%$

The correct choice is A.

For more practice with this skill go to page 237 of this manual.

CLAST SKILL IV.A.3
The student solves problems that involve the structure and logic of arithmetic.

Refer to Section 5.1 of *Thinking Mathematically*.

These are rather abstract problems that will refer to concepts such as divisibility, factorization, and remainders. We can solve these problems by making lists of numbers that satisfy stated conditions, and then comparing lists. In some cases, the problems may involve finding the least common multiple or greatest common factor of two numbers.

EXAMPLE A

Find the greatest number that is a factor of 72 and is also a divisor of 48 and a divisor of 60.

A. 18   B. 12   C. 4   D. 24

EXAMPLE A SOLUTION

Recall that the words "factor" and "divisor" mean the same thing. This means that we are asked to find the greatest common divisor of 72, 60 and 48. We can find the greatest common divisor by referring to the prime factorizations of the three numbers.

$72 = (2)(36) = (2)(2)(18) = (2)(2)(2)(9) = (2)(2)(2)(3)(3)$

$= (2^3)(3^2)$

$60 = (2)(30) = (2)(2)(15) = (2)(2)(3)(5)$

$= (2^2)(3)(5)$

$48 = (2)(24) = (2)(2)(12) = (2)(2)(2)(6) = (2)(2)(2)(2)(3)$

$= (2^4)(3)$

Using these prime factorizations we see that the greatest common factor is

$(2^2)(3) = 12$

The correct choice is B.

We could have also solved this problem by working backward from the answers.

SKILL IV.A.3

The largest answer given is 24 (choice D), but it is not correct because 24 is not a factor of 60.

The next largest answer given is 18 (choice A), but it is not correct because 18 is not a factor of 60.

The next largest answer given is 12. Since 12 is the largest answer given that is a factor of all three numbers, it must be their greatest common divisor.

EXAMPLE B

How many positive integers leave a remainder of 5 when divided into 61 and leave a remainder of 3 when divided into 45?

A. 0   B. 7   C. 2   D. 3

EXAMPLE B SOLUTION

To solve this problem we will list all of the numbers that satisfy the first condition, list all of the numbers that satisfy the second condition, and then count the numbers that appear on both lists.

There is a general fact that is useful for problems such as this:

For non-negative numbers, if a leaves a remainder of r when divided into b, then $a > r$ and a is a divisor of $b - r$.

We use this fact to list the numbers that leave a remainder of 5 when divided into 61. These are the numbers that are greater than 5 and are divisors of 56:

7, 8, 14, 28, 56

We also list the numbers that leave a remainder of 3 when divided into 45. These are the divisors of 42 that are greater than 3:

6, 7, 14, 21, 42

We see that the following numbers appear on both lists:

7, 14

The correct choice is C.

For more practice with this skill go to page 193 of this manual.

CLAST SKILL I.B.1
The student will round measurements to the nearest given unit of the measuring device.

Refer to Sections 9.1, 9.2 and 1.2 of *Thinking Mathematically*.

You will be given a measurement (that is, a number to which units of measure are attached, such as "25.38 inches" or "639.7 hours") and asked to round that measure to some specified degree of accuracy.

The units of measure involved may be from the English system (such as feet, yards, pounds, square inches, ounces, or gallons), the metric system (such as liters, meters, grams, square centimeters, or hectares) or may be units of time (such as seconds, minutes or hours).

For problems involving English units of measure, you may also be required to convert units (for instance from inches to yards).

The problems may also refer to measurements presented in diagrams.

Section 1.2 of *Thinking Mathematically* has a brief review of rounding numbers.

EXAMPLE A

Round the measure 2.4539 square centimeters to the nearest tenth of a square centimeter.

A. 2 sq cm

B. 2.45 sq cm

C. 2.4 sq cm

D. 2.5 sq cm

SKILL I.B.1

EXAMPLE A SOLUTION

To round 2.4539 square centimeters to the nearest tenth, we examine the digit in the hundredths place:

2.4<u>5</u>39

Since this digit is greater than 4, we round up, as follows:

Drop all digits to the right of the tenths place and add 1 to the digit in the tenths place.

The rounded value is 2.5 sq cm.

The correct choice is D.

EXAMPLE B

Round the measure 49.2 feet to the nearest ten yards.

A. 50 yards   B. 40 yards   C. 20 yards   D. 150 yards

EXAMPLE B SOLUTION

First we must convert units, because the measurement is given in feet but we want the answer in yards. To convert from feet to yards, we divide by 3.

$$49.2 \text{ feet} \times \frac{1 \text{ yard}}{3 \text{ feet}} = \frac{49.2}{3} \text{ yards} = 16.4 \text{ yards}$$

Now we round 16.4 yards to the nearest ten. To do so, we examine the digit in the ones position:

1<u>6</u>.4

Since this digit is greater than 4, we round up by replacing all digits to the right of the tens place with zeros and adding one to the digit in the tens place. We drop all trailing zeros to the right of the decimal point.

The correct choice is C, 20 yards.

For more practice with this skill go to pages 241 and 243 of this manual.

CLAST SKILL I.B.2
The student will calculate distances.
The student will calculate areas.
The student will calculate volumes.

CLAST SKILL I.B.2a
The student will calculate distances.

Refer to Sections 10.3 and 10.4 of *Thinking Mathematically.*

You will be asked to compute the perimeter of a polygon or the circumference of a circle. The calculation might also require the Pythagorean theorem. It may be necessary to perform English-to-English or metric-to-metric unit conversions; if unit conversions are required, you will not be given the conversion factors. Diagrams might accompany the problems.

EXAMPLE A

Find the distance around a circular region whose radius is 3 meters.

A. 9π m

B. 36π m

C. 6π m

D. 3π m

EXAMPLE A SOLUTION

This problem refers to the formula for the circumference of a circle: $C = 2\pi r = \pi D$.

In this case, $r = 3$, so $C = 2\pi(3) = 6\pi$ m.

The correct choice is C.

SKILL I.B.2

EXAMPLE B

Find the distance around the regular hexagon shown below.

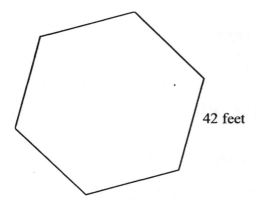

42 feet

A. 252 yards

B. 84 yards

C. 252 square yards

D. 84 square yards

EXAMPLE B SOLUTION

Since this is a *regular* hexagon, all six sides have the same length (42 feet).

However, the answers are all expressed in terms of yards rather than feet, so we will convert the measurement to yards before be calculate the perimeter.

42 feet $42 \text{ ft} = 42 \text{ ft} \times \dfrac{1 \text{ yd}}{3 \text{ ft}} = \dfrac{42}{3} \text{ yd} = 14 \text{ yd}$

The total distance around all 6 sides is then 6(14) = 84 yd.

The correct choice is B.

Choice D has the wrong units (distances are measured in linear units, not square units).

For more practice with this skill go to page 266 of this manual.

CLAST SKILL I.B.2b
The student will calculate areas.

Refer to Section 10.4 of *Thinking Mathematically.*

You will be asked to compute the area of one of these figures:
circle, square, rectangle, triangle, parallelogram. You might also be required to find the surface area of a rectangular solid. It may be necessary to perform English-to-English or metric-to-metric unit conversions; if unit conversions are required, you will not be given the conversion factors. Diagrams might accompany the problems.

EXAMPLE C

Find the area of a circular garden with a diameter of 10 feet.

A. 25π sq. ft

B. 10π sq. ft

C. 100π sq. ft

D. 50π sq. ft

EXAMPLE C SOLUTION

This problem refers to the formula for the area of a circle: $A = \pi r^2$.

In this case, the diameter is 10 ft, so r = 5 ft.

$A = \pi 5^2 = 25\pi$ sq. ft.

The correct choice is A.

SKILL I.B.2

EXAMPLE D

Find the area of the figure shown below.

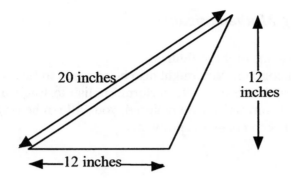

A. 72 sq. ft

B. 12 sq. ft

C. 5 sq. ft

D. $\frac{1}{2}$ sq. ft

EXAMPLE D SOLUTION

This refers to the formula for the area of a triangle: $A = \frac{1}{2}bh$

In this case, b = 12 inches and h = 12 inches (the 20-inch side is not a factor in this calculation).

However, the answers are expressed in terms of square feet, not square inches, so before we use the formula for area we will convert the measurements to feet:

b = 1 ft, h = 1 ft

Now, we have

$A = \frac{1}{2}(1 \text{ ft})(1 \text{ ft}) = \frac{1}{2}$ sq. ft

The correct choice is D.

For more practice with this skill go to pages 268 and 281 of this manual.

CLAST SKILL I.B.2c
The student will calculate volumes.

Refer to Section 10.5 of *Thinking Mathematically.*

You will be asked to compute the volume of one of these figures:
sphere, rectangular solid, right circular cone, right circular cylinder. It may be necessary to perform English-to-English or metric-to-metric unit conversions; if unit conversions are required, you will not be given the conversion factors. Diagrams might accompany the problems.

EXAMPLE E

Find the volume of the cone shown below.

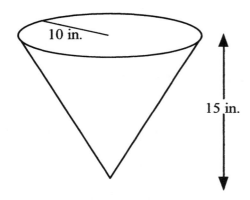

A. 1500π cu. in

B. 500π cu. in

C. 750π cu. in

D. 2250π cu. in

EXAMPLE E SOLUTION

This problem refers to the formula for the volume of a right circular cone: $V = \frac{1}{3}\pi r^2 h$.

In this case, r = 10 in and h = 15 in, so

SKILL I.B.2

$$V = \frac{1}{3}\pi\left(10^2\right)(15)$$

$$= \frac{1}{3}\pi\left(10^2\right)(15)$$

$$= \frac{1}{3}\pi\left(10^2\right)(15)$$

$$= \frac{1}{3}\pi(100)(15)$$

$$= 500\pi \text{ cu. in}$$

The correct choice is B.

EXAMPLE F

Find the volume of a rectangular solid measuring 6 feet long, 3 feet wide and 9 inches high.

A. 162 cu. ft   B. 1944 cu. ft   C.   162 cu, ft   D. 13.5 cu. ft

EXAMPLE F SOLUTION

This refers to the formula for the volume of a rectangular solid: $V = LWH$.

In this case, the answers are expressed in terms of cubic feet, but one of the measurements is given in inches. We will convert this measurement to feet before using the formula for volume.

$H = 9$ inches

$$= 9 \text{ in} \times \frac{1 \text{ ft}}{12 \text{ in}} = \frac{9}{12} \text{ ft} = \frac{3}{4} \text{ ft} = .75 \text{ ft}$$

Now we compute the volume.

$V = LWH = (6 \text{ feet})(3 \text{ feet})(.75 \text{ feet}) = 13.5$ cu. ft.

The correct choice is D.

For more practice with this skill go to page 281 of this manual.

CLAST SKILL II.B.1
The student will identify relationships between angle measures.

Refer to Sections 10.1, 10.2, and 10.3 of *Thinking Mathematically*.

This skill will test your ability to perform computations involving facts about vertical angles, corresponding angles, alternate interior angles, and alternate exterior angles. The problems may also require that you recognize and use facts about triangles, including isosceles triangles and equilateral triangles.

EXAMPLE A
Given that $\overline{AC} \parallel \overline{BD}$, which of the following statements is true for the figure shown? (The measure of angle ACD is represented by "x.")

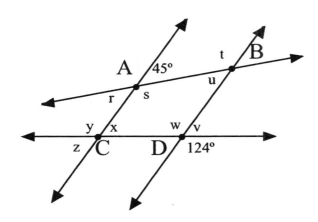

A. $y = s$ B. $t = 56°$ C. $u = 45°$ D. $v = u$

EXAMPLE A SOLUTION

First, recall the meaning of the notation.

You need to be aware that "$\overline{AC} \parallel \overline{BD}$" means "the line segment connecting vertex A to vertex C is parallel to the line segment connecting vertex B to vertex D."

Based on the 124° angle at vertex D, we obtain the following result:

Because of supplementary angles, $v = 56°$.

Next, because of vertical angles, corresponding angles, alternate interior angles or alternate exterior angles,

37

SKILL II.B.1

w = 124°

y = 124°

x = 56°

z = 56°

Based on the 45° angle at vertex A, we have this result:

s = 135° because it is the measure of an angle supplementary to the given 45° angle.

Using facts about vertical angles, corresponding angles, alternate interior angles or alternate exterior angles, we obtain the following results:

r = 45°

t = 135°

u = 45°

We see that the correct choice is C.

EXAMPLE B

Given that $\overline{AB}\perp\overline{BC}$ and $\overline{AB}\cong\overline{BC}$, which of the following statements is true for the figure shown?

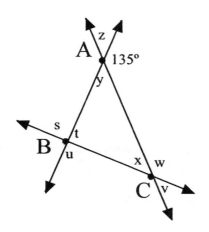

A. △ABC is equilateral   B. t = 135°   C. x = 55°   D. v = z

EXAMPLE B SOLUTION

First, recall the meaning of the notation.

You need to be aware that "$\overline{AB} \perp \overline{BC}$" means "the line segment connecting vertex A to vertex B is *perpendicular* to the line segment connecting vertex B to vertex C."

You also need to be aware that "$\overline{AB} \cong \overline{BC}$" means "the line segment connecting vertex A to vertex B is *congruent* to the line segment connecting vertex B to vertex C."

Congruent figures are identical. For instance, congruent line segments have the same length, congruent angles have the same measure, congruent triangles have the same shape and size, and so on. If two figures are congruent, then they "can be made to coincide" in the words of Euclid; that is, one of the figures can be placed over the other figure in such a way that they match each other exactly.

In this case, we are being told that the two line segments are exactly the same length.

Due to supplementary angles, we know that y = 45° and z = 45°.

Because $\overline{AB} \cong \overline{BC}$, we know that $\triangle ABC$ is isosceles.

Since in an isosceles triangle the angles opposite the equal sides are congruent, we know that y = x; that is, y = 45°.

Finally, due to vertical angles we know that v = 45°.

Since v = 45° and z = 45°, we see that the correct choice is D.

For more practice with this skill go to pages 246 and 251 of this manual.

CLAST SKILL II.B.2
The student will classify simple plane figures by recognizing their properties.

Refer to Sections 10.1, 10.2, and 10.3 of *Thinking Mathematically*.

This CLAST skill is primarily a test of your understanding of nomenclature involving angles, triangles and quadrilaterals.

You will be shown a geometric figure or given a verbal description of the properties of a geometric object and asked to identify the correct geometric object or the portion of the figure corresponding to the specified geometric object.

The problem may involve the terminology used to classify angles or angle pairs, such as *acute, right, obtuse, vertical, supplementary, complementary, alternate interior, alternate exterior,* and *corresponding*.

The problem may also involve the terminology used to classify triangles, such as *acute, scalene, obtuse,* and *right*.

Finally, the problem may involve the terminology used to classify convex quadrilaterals, limited to *square, rectangle, parallelogram, rhombus,* and *trapezoid*.

EXAMPLE A

In the figure below, $\overline{AB} \parallel \overline{CD}$.
Which of the following is a pair of corresponding angles?

A. 1 and 2   B. 5 and 6   C. 3 and 2   D. 5 and 7

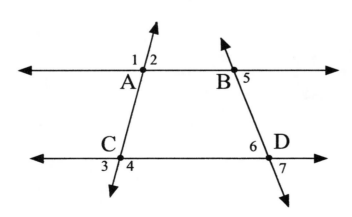

EXAMPLE A SOLUTION

Recall when two parallel lines are crossed by a transversal, corresponding angles occupy the same position relative to the transversal and one of the parallel lines.

In this figure, angles 5 and 7 are corresponding angles; this is the only pair of corresponding angles among the seven numbered angles in the figure.

The correct choice is D.

For further practice, verify that each of the following is true:

Angles 2 and 3 are a pair of alternate exterior angles, as are angles 1 and 4.
Angles 1 and 2 are a pair of supplementary angles, as are angles 3 and 4.
Angles 5 and 6 are a pair of alternate interior angles.
Angles 6 and 7 are a pair of vertical angles.
Angle 2 is an acute angle, as are angles 3, 5, 6 and 7.
Angle 4 is an obtuse angle, as is angle 1.

EXAMPLE B

Select the figure that can simultaneously possess <u>all</u> of the following characteristics:

i.   is a quadrilateral;
ii. has no right angles;
iii. opposite angles are congruent;
iv. all sides have the same length.

A. rectangle   B. equilateral triangle   C. rhombus   D. square

EXAMPLE B SOLUTION

We can eliminate the incorrect choices.

Choice B is not correct because the figure must be a quadrilateral (a four-sided figure), not a triangle.

Choices A and D are not correct because we want a figure that can have no right angles.

This leaves choice C. Recall that a rhombus is a parallelogram on which all sides have equal lengths. Among the choices listed, this is the only one that can simultaneously possess all four of the given traits.

The correct choice is C.

For more practice with this skill go to pages 245, 249, and 273 of this manual.

CLAST SKILL II.B.3
The student will recognize similar triangles and their properties.

Refer to Section 10.2 of *Thinking Mathematically.*

In one format you will be given a diagram showing similar triangles and asked to select a statement that correctly describes a relationship depicted by the diagram. The relationship may involve side lengths or angles.

In another format you will be given four diagrams, each of which shows two or more triangles, and asked to select the diagram in which all of the triangles are similar.

Recall the following **facts about similar triangles:**

1. To demonstrate that two triangles are similar, it suffices to show that they have two angles in common.

2. When triangles are similar, ratios of lengths of corresponding sides are equal.

EXAMPLE A

Which of the statements A – D is true for the pictured triangles?

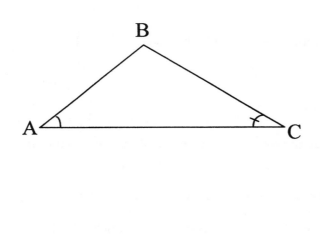 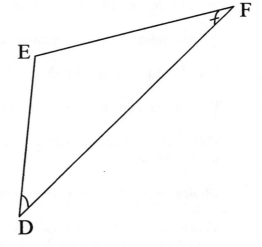

A. $\angle ABC \cong \angle EFD$　　　　B. $AB = DE$

C. $\dfrac{BC}{EF} = \dfrac{DE}{BA}$　　　　D. $\dfrac{EF}{ED} = \dfrac{BC}{BA}$

EXAMPLE A SOLUTION

First, a word or two about notation:

One way to denote or name a side of a triangle or a line segment in general is to list the two vertices joined by the line segment, under a bar.

For instance, the notation $\overline{AB}$ denotes the side of triangle ABC joining vertex A to vertex B. Also, $\overline{AB}$ is the same as $\overline{BA}$.

By contrast, the notation AB denotes the **length** of side $\overline{AB}$. Of course, AB is no different from BA.

Congruence is denoted by the symbol "$\cong$." Two geometric figures are congruent if they are equal. For instance, two angles are congruent if they have the same degree measure, two line segments are congruent if they have the same length, and two triangles are congruent if they have the same shape and the same size. In the words of Euclid, congruent figures "can be made to coincide."

On the CLAST, arcs drawn in angles may be used to denote congruence. In the figure above, the fact that angles BAC and EDF each have an arc drawn through them indicates that those two angles are congruent. Likewise, the fact that angles EFD and BCA each contain an arc with a slash indicates that those two angles are congruent.

The fact that triangle ABC has two angles in common with triangle DEF is sufficient to tell us that the triangles are similar.

Once we know that two triangles are similar, there are a great many statements that will correctly describe relationships between components of the triangles. In order to quickly distinguish correct statements from incorrect statements, it is useful to construct a **correspondence table**. We will do this for the triangle shown above. First note that $\overline{AC}$ on the first triangle corresponds to $\overline{DF}$ on the second triangle, because in each case the side connects the angle having the arc to the angle having the arc and slash. For similar reasons, $\overline{AB}$ corresponds to $\overline{DE}$ and $\overline{BC}$ corresponds to $\overline{EF}$.

These crucial observations are summarized in the table below.

side of △ABC	corresponding side of △DEF
$\overline{AC}$	$\overline{DF}$
$\overline{AB}$	$\overline{DE}$
$\overline{BC}$	$\overline{EF}$

By referring to this table we can instantly form correct statements of proportion.

For instance, suppose we want to form a correct proportion involving $\dfrac{AC}{AB}$.

Since this fraction involves quantities associated with the first two cells in the left column, it will be equal to the fraction formed by using the quantities associated with the corresponding two cells in the right column:

$$\frac{AC}{AB} = \frac{DF}{DE}$$

Following this approach, we can make this correct proportion as well:

$$\frac{EF}{DE} = \frac{BC}{AB}$$

Notice that in the previous proportion, the numerator and denominator in the first fraction are associated with the third and second cells, respectively, of the right-hand column of the table, while the numerator and denominator of the second fraction are associated with the corresponding two cells from the left-hand column.

Here are some other correct proportions that can be formed or verified in a similar way:

$$\frac{BA}{BC} = \frac{DE}{FE}$$

$$\frac{FD}{EF} = \frac{CA}{CB}$$

Another way to form a correct proportion is illustrated in this case:

$$\frac{AC}{DF} = \frac{AB}{DE}$$

Notice that this proportion was formed in a slightly different manner from the others. In this case, the numerator of the first fraction is associated with the first cell in the left-hand column, while the denominator of the first fraction is associated with the first cell in the right-hand column; in the second fraction, the numerator and denominator are associated with the second cells of the left-hand and right-hand columns, respectively. This is an illustration of first working "horizontally" rather than "vertically" in the table in order to find or verify correct proportions. Here are some other correct proportions that can be formed in this manner:

$$\frac{EF}{CB} = \frac{ED}{AB} \qquad \frac{BA}{DE} = \frac{AC}{DF} \qquad \frac{DF}{AC} = \frac{EF}{BC}$$

Referring to the multiple-choice options given in this sample problem, we see that the correct response is D.

Verify that choice C is incorrect because the fraction on the right-hand side of the equals sign is "upside down."

Choice B is incorrect because it states that two side lengths are equal. Although the two associated sides are corresponding sides, that does not mean that they must have the same length. In this case it is easy to see that $\overline{DE}$ is longer than $\overline{AB}$.

Choice A is incorrect because the arc notation in the figure tells us that angle ABC is congruent to angle DEF, not to angle EFD.

EXAMPLE B

Which of the statements A – D is true for the pictured triangles?

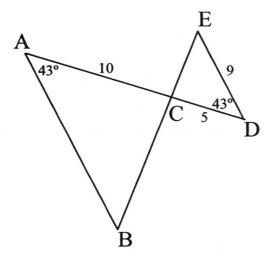

A. $\dfrac{AB}{ED} = \dfrac{AC}{EC}$ B. $\dfrac{10}{AB} = \dfrac{5}{9}$

C. $\dfrac{EC}{CD} = \dfrac{CB}{AB}$ D. $AB = 45$

EXAMPLE B SOLUTION

First we verify that the triangles are similar.

We are told that angle CAB is congruent to angle CDE, and the properties of vertical angles tell us that angle ACB is congruent to angle ECD. Since the triangles have two angles in common, they are similar.

Next we make a table of corresponding sides. To correctly associate corresponding sides we first identify corresponding vertices. Vertex A on triangle ABC corresponds to vertex

D on triangle EDC (because they both occur at the 43° angles), vertex C on triangle ABC corresponds to vertex C on triangle EDC, and likewise vertex B corresponds to vertex E.

side of $\triangle$ABC	corresponding side of $\triangle$DEF
$\overline{AB}$	$\overline{DE}$
$\overline{BC}$	$\overline{EC}$
$\overline{AC}$	$\overline{DC}$

Based on this table we see, for instance, that

$$\frac{AB}{ED} = \frac{BC}{EC} \text{ and } \frac{AB}{ED} = \frac{AC}{DC}$$

The statement in choice A is incorrect.

Likewise, the table shows that

$$\frac{AC}{AB} = \frac{DC}{DE}$$

Substituting the given values 10, 5 and 9 for AC, DE and DC respectively, we see that choice B is correct.

Choice C is incorrect because $\dfrac{EC}{CD} = \dfrac{CB}{AC}$.

As for choice D, to correctly find AB we can use this relationship:

$$\frac{AB}{AC} = \frac{ED}{CD}$$

Substituting given values for AC, ED and CD we have

$$\frac{AB}{10} = \frac{9}{5}$$

Solving for AB we have

$$AB = \frac{10 \times 9}{5} = 18 \qquad \text{The correct choice is B.}$$

For more practice with this skill go to page 255 of this manual.

CLAST SKILL II.B.4
The student identifies appropriate units of measure for geometric objects.

Refer to Sections 9.1 and 9.2 of *Thinking Mathematically*.

This skill tests your ability to distinguish between measurements of **distance, area** and **volume.** You will not be required to use any geometric formulas or to perform any computations in solving these problems.

Recall the following basic facts.

Distances, such as diameters, heights, or side lengths of geometric figures, are measured in **linear units**, like inches, feet, yards, miles, millimeters, centimeters, meters, or kilometers.

Areas, which describe the amount of two-dimensional space occupied by a flat object or the amount of flat material required to cover a three-dimensional object, are measured in **square units,** such as square inches, square feet, square yards, square miles, square millimeters, square centimeters, square meters, or square kilometers. Acres and hectares are also units of square measure, although those names don't contain the word "square."

Volumes, which describe the amount of space occupied by a three-dimensional object or the capacity of a three-dimensional container, are measured in **cubic units** such as cubic inches, cubic feet, cubic yards, cubic miles, cubic millimeters, cubic centimeters, cubic meters, or cubic kilometers. Gallons, quarts, and liters are other examples of cubic measure, although those names don't contain the word "cubic."

EXAMPLE A

Identify the units of measure that would be appropriate for measuring the amount of flat linoleum needed to cover a rectangular floor.

A. feet

B. yards

C. square inches

D. cubic centimeters

SKILL II.B.4

EXAMPLE A SOLUTION

In this case we are measuring the **area** of the floor.

Areas are measured with square units (not linear or cubic units).

The only appropriate choice is C.

EXAMPLE B

Identify the units of measure that would be appropriate for measuring the amount of water contained in a cylindrical fish tank.

A. meters
B. centimeters
C. square millimeters
D. cubic centimeters

EXAMPLE B SOLUTION

In this case we are measuring the capacity or **volume** of the three-dimensional vessel.

Volumes are measured with cubic units (not linear or square units).

The only appropriate choice is D.

EXAMPLE C

Identify the units of measure that would be appropriate for measuring the diameter of a spherical stone.

A. inches
B. square feet
C. cubic yards
D. square meters

EXAMPLE C SOLUTION

In this case we are measuring a **distance.**

Distances are measured with linear units (not cubic or square units).

The only appropriate choice is A.

For more practice with this skill go to page 243 of this manual.

CLAST SKILL III.B.1
The student infers formulas for measuring geometric figures.

Refer to Sections 10.1, 10.3, 10.4, and 10.5 of *Thinking Mathematically.*

The point of this CLAST skill is to challenge you to discover or infer a geometric formula.

You will be shown a series of geometric figures. The figures will not necessarily depict familiar objects or refer directly to concepts that you have encountered in your studies of geometry. Along with each figure there will be numbers associated with calculations involving measurements such as distances, areas or volumes. You will be asked to perform a related calculation.

Your goal is to study the numbers that accompany the given figures, trying to recognize a pattern that can be used to solve the given problem. *If you fail to realize that the objective is to recognize and use a pattern, these problems may seem extraordinarily difficult.*

EXAMPLE A
Study the given information. In each figure, "h" represents the height of the figure.

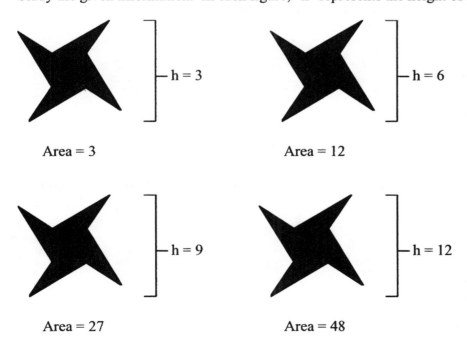

Area = 3   Area = 12

Area = 27   Area = 48

Calculate the area of a similar figure if the height is 18.
A. 108   B. 72   C. 36   D. 54

EXAMPLE A SOLUTION

First we will transfer the data from the figures into a table:

h	area
3	3
6	12
9	27
12	48

Notice that the data in the first two rows suggest that if h is doubled, area is increased by a factor of four.

Comparing the data in the second row with the data in the last row, again the numbers suggest that if h is doubled, area is quadrupled.

This observation allows us to use the data in the third row to find the area if h = 18. Since area = 27 when h = 9, we expect that when h = 18 area = (27)(4) = 108.

The correct choice is A.

Another way to infer a useful pattern is to look for common factors in the numbers. Notice that all of the numbers in the table are multiples of 3. We will rewrite the numbers in the table as products involving factors of 3.

h	area
3(1)	3(1)
3(2)	3(4)
3(3)	3(9)
3(4)	3(16)

Now notice that all of the numbers in the "area" column have perfect square factors.

h	area
3(1)	$3(1^2)$
3(2)	$3(2^2)$
3(3)	$3(3^2)$
3(4)	$3(4^2)$

Now we see a definite pattern in the numbers:

h	area
3(1)	$3(1^2)$
3(2)	$3(2^2)$
3(3)	$3(3^2)$
3(4)	$3(4^2)$
3(n)	$3(n^2)$

Applying this pattern to the original problem, when h = 18 (=3(6)), area = $3(6^2)$ = 108.

Again we see that the correct choice is A.

General hint: in order to infer a formula involving areas, see if it is possible to rewrite areas in factored form involving perfect square factors.

EXAMPLE B

Referring to the information in the previous problem, find the area of a similar figure if the height is 5.

A. 75   B. $\dfrac{9}{5}$ C. $\dfrac{25}{3}$ D. 9

EXAMPLE B SOLUTION

We have already found a formula to use: if h = 3n, then area = $3n^2$.

In order to use this formula, however, we must be able to write 5 as a number of the form 3n. We can do this using fractions: 5 = 3(5/3).

Thus, when h = 5, area = $3\left(\dfrac{5}{3}\right)^2 = 3\left(\dfrac{25}{9}\right) = \dfrac{25}{3}$

The correct choice is C.

EXAMPLE C

Study the three-dimensional figures shown below. In each figure, S represents the length of a side or edge of the figure, and V represents the volume.

SKILL III.B.1

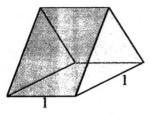

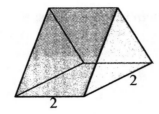

S = 1
V = $\sqrt{3}$

S = 2
V = $8\sqrt{3}$

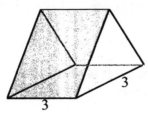

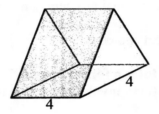

S = 3
V = $27\sqrt{3}$

S = 4
V = $64\sqrt{3}$

Find the volume of a similar figure if S = 6.

A. $54\sqrt{3}$

B. $216\sqrt{3}$

C. $68\sqrt{3}$

D. $91\sqrt{3}$

EXAMPLE C SOLUTION

First we use a table to summarize the numbers in the figures.

S	V
1	$\sqrt{3}$
2	$8\sqrt{3}$
3	$27\sqrt{3}$
4	$64\sqrt{3}$

When we compare the data in the first row with the data in the second row, the inference is that if S is doubled, V is multiplied by a factor of 8. If we compare the data in the second row with the data in the fourth row, again the inference is that if S is doubled V is multiplied by a factor of 8. This suggests that to find V when S = 6, we use the data in the third row: since $V = 27\sqrt{3}$ when S = 3, when S = 6, $V = 8\left(27\sqrt{3}\right) = 216\sqrt{3}$.

The correct choice is B.

Another way to infer a pattern is by comparing the numbers in the second column of the table. Notice that each of those numbers has a perfect cube factor:

S	V
1	$\left(1^3\right)\sqrt{3}$
2	$\left(2^3\right)\sqrt{3}$
3	$\left(3^3\right)\sqrt{3}$
4	$\left(4^3\right)\sqrt{3}$
n	$\left(n^3\right)\sqrt{3}$

Applying this formula to the case in which n = 6, we have $V = \left(6^3\right)\sqrt{3} = 216\sqrt{3}$.

Again, we see that the correct choice is B.

General hint: in order to infer a formula involving volumes, see if it is possible to rewrite volumes in factored form involving perfect cube factors.

For more practice with this skill go to pages 263, 274, and 285 of this manual.

CLAST SKILL III.B.2
The student identifies applicable formulas for computing measures of geometric figures.

Refer to Sections 10.3, 10.4, and 10.5 of *Thinking Mathematically.*

You will be shown a composite geometric figure that consists of two or three adjacent simpler figures. Parts of the composite figure, such as side lengths, will be represented by symbols. You will be asked to select a formula describing the perimeter, area, volume, surface area, or some other attribute of the composite figure. The formula will involve the same symbols that were used to label parts of the composite figure.

Two-dimensional figures used in these problems may include rectangles, squares, parallelograms, rhombuses, circles, semi-circles, and triangles. You need to know the general formulas for calculating the areas of these figures. Three-dimensional figures may include rectangular solids, cylinders, and right circular cones. You need to know the formulas for the volumes of these figures, as well as the surface area of the rectangular solid.

EXAMPLE A

Study the figure below, consisting of two triangles. The sides whose lengths are b and B respectively are parallel.

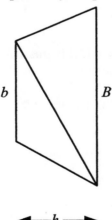

Select the formula for calculating the area of the figure.

A. Area $= \dfrac{1}{2}bh + \dfrac{1}{2}Bh$

B. Area $= \dfrac{1}{2}bBh$

C. Area $= \dfrac{1}{2}b^2 + \dfrac{1}{2}B^2 + \dfrac{1}{2}h^2$

D. Area $= \dfrac{1}{2}b + \dfrac{1}{2}B + \dfrac{1}{2}h$

EXAMPLE A SOLUTION

Since the figure consists of two triangles, we make this observation:

Area of figure = area of left-hand triangle + area of right-hand triangle.

Now, for the left-hand triangle, b is the length of the base, and h is the height, so

Area of left-hand triangle = $(1/2)bh$.

For the right-hand triangle, the length of the base is B and the height is again h, so

Area of right-hand triangle = $(1/2)Bh$.

Thus,

Area of figure = $(1/2)bh + (1/2)Bh$.

The correct choice is A.

SKILL III.B.2

EXAMPLE B

Study the figure below, showing a right circular cone attached to a right circular cylinder.

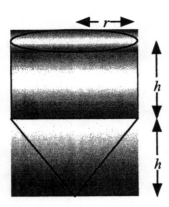

Select the formula for computing the volume of the figure.

A. Volume $= \dfrac{2}{3}\pi r^2(2h)$

B. Volume $= \pi r^2(2h) + \dfrac{1}{3}\pi r^2(2h)$

C. Volume $= \dfrac{2}{3}\pi r^2 h$

D. Volume $= \pi r^2 h + \dfrac{1}{3}\pi r^2 h$

EXAMPLE B SOLUTION

We make this observation:

Volume of figure = volume of cylinder + volume of cone.

According to the labels in the figure, r is the radius of the circular bases of both the cone and cylinder, and h is the height of both the cone and the cylinder.

Volume of cylinder $= \pi r^2 h$.   Volume of cone $= \dfrac{1}{3}\pi r^2 h$.

So, Volume of figure $= \pi r^2 h + \dfrac{1}{3}\pi r^2 h$

The correct choice is D.

For more practice with this skill go to pages 278 and 290 of this manual.

CLAST SKILL IV.B.1
The student solves real world problems involving perimeters, areas and volumes of geometric figures.

Refer to Sections 10.3, 10.4 and 10.5 of *Thinking Mathematically.*

You will be given a real-world problem which involves the calculation of the perimeter, area or volume of a geometric figure. The problem will be taken from a business, social studies, industry, education, economics, environmental studies, arts, physical science, sports or consumer-related context. Solving the problem should not require specialized knowledge of those areas, however. A diagram may accompany the problem.

Conversion of units will be required.

EXAMPLE A

The figure below shows a region of a wilderness area in which a hiker has been reported missing. Seachers estimate that they can cover 1 million square meters per day. At this rate, how many days will be required to search the entire region?

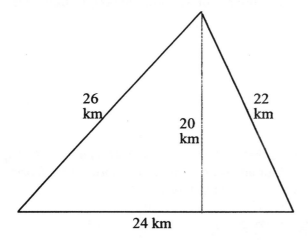

A. .24 days

B. .072 days

C. 13.73 days

D. 240 days

SKILL IV.B.1

EXAMPLE A SOLUTION

To solve this problem, we need to find the area of the triangular region (in square meters) and divide by the rate of 1 million square meters per day.

The formula for the area (A) of a triangle is $A = \frac{1}{2}bh$.

In this figure, b = 24 km and h = 20 km.

Since we want the area to be expressed in square meters, rather than square kilometers, we will convert the measurements to meters before using the formula for area.

Since there are 1000 meters per kilometer, we have

b = 24,000 m

h = 20,000 m

Then A = (1/2)(24,000 m)(20,000 m) = 240,000,000 sq m.

The number of days required to search the entire region is (240,000,000)/(1,000,000) = 240

The correct choice is D.

EXAMPLE B

The diagram below shows a piece a canvas that will be used to construct a sail. A strip of reinforcing fabric will be sewn along all four edges of the piece of fabric. This fabric costs 5¢ per inch. Find the total cost of the reinforcing fabric.

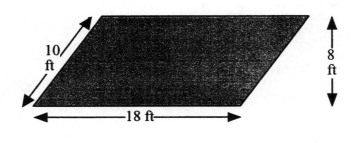

A. $31.20   B. $33.60

C. $2.80   D. $7.20

58

EXAMPLE B SOLUTION

To solve this problem we need to find the perimeter of the figure (in inches) and multiply by the cost factor of 5¢ per inch.

The measurements in the figure are given in feet, but we want to find the perimeter in inches, so we will convert the pertinent measurements from feet to inches by multiplying by 12.

$$10 \text{ ft} = 10 \text{ ft} \times \frac{12 \text{ in}}{1 \text{ ft}} = 120 \text{ in}$$

$$18 \text{ ft} = 18 \text{ ft} \times \frac{12 \text{ in}}{1 \text{ ft}} = 216 \text{ in}$$

The perimeter of the figure is then

$$120 + 120 + 216 + 216 = 672 \text{ in.}$$

The total cost of the reinforcing fabric is (672 in)($.05 per in) = $33.60.

The correct choice is B.

EXAMPLE C

The figure below shows the outline of the foundation of a house. The floor will be a concrete slab that is 6 inches thick. If concrete costs $80 per cubic yard, find the cost of the concrete for this job.

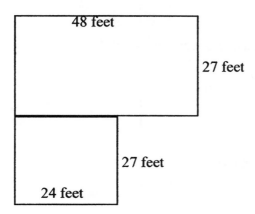

A. $145.80   B. $2880   C. $960   D. $320

SKILL IV.B.1

EXAMPLE C SOLUTION

We need to find the volume (in cubic yards) of this solid object, and then multiply by the cost factor of $80 per cubic yard.
Since we want the volume expressed in cubic yards, we will convert all measurements to yards before using the formula for volume.

To convert from feet to yards, we divide by 3.

48 feet = 16 yards   24 feet = 8 yards

To convert from inches to yards, we divide by 36.

6 inches = 6/36 yard = 1/6 yard

Now we consider the foundation slab to consist of two rectangular solids:

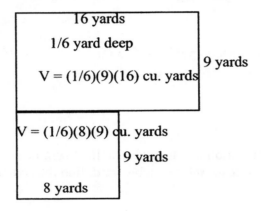

As the diagram indicates, the total volume of the two rectangular solids is

$(16)(9)(1/6) + (9)(8)(1/6)$ cu. yds, or $\dfrac{144 + 72}{6}$ cu. yd

This reduces to 36 cu. yd.

The total cost is (36 cu. yd)($80 per cu. yd) = $2,880

The correct choice is B.

For more practice with this skill go to pages 270 and 283 of this manual.

CLAST SKILL IV.B.2
The student solves real-world problems involving the Pythagorean property.

Refer to Section 10.2 of *Thinking Mathematically*.

You will be given a real-world problem involving a situation from business, social studies, industry, education, economics, environmental studies, the arts, physical science, sports, or a consumer-related context. The problem will not require specialized knowledge of any of those fields, however.

The problem will be appropriate for application of the Pythagorean theorem. You may be required to convert units within the English system or within the metric system.

The measures of the sides of the right triangle will be whole numbers less than or equal to 16. This means that the right triangle will be limited to one of these configurations: 3-4-5, 6-8-10, 9-12-15, or 5-12-13 (recall, for instance, that 3-4-5 refers to a right triangle where one leg measures 3 units, the other leg measures 4 units, and the hypotenuse measures 5 units).

EXAMPLE A

The figure below shows the route that Andre travels 5 days per week. First he goes from home to school. After classes are over he travels from school to work, and at the end of his shift he returns home. Assuming that he takes no side trips, how many miles does he travel each day?

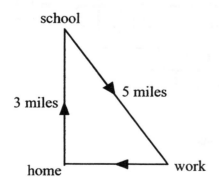

A. 4 miles   B. 5.8 miles   C. 13.8 miles   D. 12 miles

SKILL IV.B.2

EXAMPLE A SOLUTION

The total distance traveled is the distance from home to school (3 miles) plus the distance from school to work (5 miles) plus the distance from work to home (x miles).
To find x, we use the Pythagorean theorem, bearing in mind that x is the length of a leg (not the hypotenuse) of a right triangle.

$5^2 = x^2 + 3^2$

$25 = x^2 + 9$

$25 - 9 = x^2$

$16 = x^2$

$\sqrt{16} = x$

x = 4 miles

The total distance is 3 + 5 + 4 = 12 miles.   The correct choice is D.

EXAMPLE A
The map below shows the region of Central City near Lake Eloise. The city council is going to build a by-pass connecting 8th Street to 6th Avenue as shown in the map. The city blocks pictured each measure 100 yards by 100 yards. It is estimated that construction of this road will cost $100 per foot. Find the total cost.

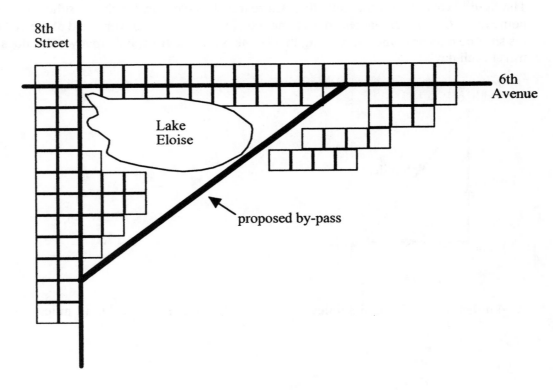

A. $450,000   B. $150,000

C. $1,000,000   D. $100,000

EXAMPLE A SOLUTION

The length of the proposed roadway can be found by determining the length of the hypotenuse of a right triangle. One leg of the right triangle measures 9 city blocks, and the other leg measures 12 city blocks, so the length of the hypotenuse (L) is given by

$$L^2 = 9^2 + 12^2$$

$$L^2 = 81 + 144$$

$$L^2 = 225$$

$$L = \sqrt{225}$$

L = 15 blocks

Since one block is 100 yards long, L = (15)(100) = 1,500 yards

The cost information is given in terms of feet, not yards, so we convert this distance to feet. To convert from yards to feet we multiply by 3

L = 1,500 yards = (1,500 yards)(3 feet/yard) = 4,500 feet

Now apply the cost factor:

Cost = (4,500 feet)($100/foot) = $450,000

The correct choice is A.

Note that this problem involved a 9-12-15 right triangle.

For more practice with this skill go to page 260 of this manual.

CLAST SKILL I.C.1a   *The student will add and subtract real numbers.*

Refer to Section 5.4 of Thinking Mathematically.

This skill is primarily a test of your understanding of simplification of expressions involving square root radicals. You will be given an expression involving two or three numbers, some of which may be irrational numbers. The irrational numbers may involve a factor of π, or may involve square root radicals. You will have to perform up to two operations of addition or subtraction, and simplification of radical expressions may be required prior to or after the additions or subtractions.

Remember: In order to add or subtract radical expressions, they must be *like* expressions; that is, they must have exactly the same radical parts:

$$a\sqrt{s} \pm b\sqrt{s} = (a \pm b)\sqrt{s}$$

EXAMPLE A

$\sqrt{27} - \sqrt{3} =$

A. 6   B. $2\sqrt{6}$ C. $2\sqrt{3}$ D. $-\sqrt{6}$

SOLUTION

In order to combine these two expressions by subtraction they must have exactly the same radical parts.

We determine if they have the same radical parts be writing them in simplest radical form.

We can simplify $\sqrt{27}$, because 27 has a perfect square factor (9).

$$\sqrt{27} = \sqrt{9 \cdot 3} = \sqrt{9} \cdot \sqrt{3} = 3\sqrt{3}$$

So,

$$\sqrt{27} - \sqrt{3} = 3\sqrt{3} - \sqrt{3} = (3 - 1)\sqrt{3} = 2\sqrt{3}$$

The correct choice is C.

For more practice with this skill go to page 201 of this manual.

CLAST SKILL I.C.1b
The student will multiply and divide real numbers.

Refer to Section 5.4 of Thinking Mathematically.

Again, this skill is primarily a test of your understanding of the process of simplifying and combining square root radical expressions. You will be given an expression involving multiplication or division of up to three factors, some of which may be irrational numbers. After performing the multiplication or division, you may need to simplify the product or quotient.

Remember:
To multiply two square root radical expressions, you multiply the radicands:

$$\sqrt{a} \cdot \sqrt{b} = \sqrt{a \cdot b}$$

To divide a number by a square root radical expression, you rationalize the denominator:

$$\frac{a}{\sqrt{b}} = \frac{a\sqrt{b}}{\sqrt{b} \cdot \sqrt{b}} = \frac{a\sqrt{b}}{b}$$

EXAMPLE B

$$\frac{\sqrt{6}}{\sqrt{10}} = \qquad A. \ \frac{\sqrt{15}}{5} \qquad B. \ 5\sqrt{3} \qquad C. \ \frac{5\sqrt{3}}{3} \qquad D. \ \frac{\sqrt{15}}{50}$$

SOLUTION

We will rationalize the denominator by multiplying the numerator and denominator by $\sqrt{10}$.

$$\frac{\sqrt{6}}{\sqrt{10}} = \frac{\sqrt{6} \cdot \sqrt{10}}{\sqrt{10} \cdot \sqrt{10}} = \frac{\sqrt{60}}{10}$$

Now simplify numerator and denominator.

We can simplify $\sqrt{60}$ because 60 has a perfect square factor.

$$= \frac{\sqrt{4} \cdot \sqrt{15}}{10} = \frac{2\sqrt{15}}{10}$$

We can reduce the fraction by canceling the common factor of 2.

$$= \frac{\sqrt{15}}{5}$$

The correct choice is A.

For more practice with this skill go to page 201 of this manual.

CLAST SKILL I.C.2
The student will apply the order of operations agreement to computations involving numbers and variables.

Refer to Sections 5.2, 5.3, and 6.1 of *Thinking Mathematically.*

You will be given an expression involving numbers or variables. The solution will be obtained by using the order of operations agreement to simplify the given expression.

Remember: The slogan "Please Excuse My Dear Aunt Sally" is used to recall the correct order in which operations are performed: Parentheses, Exponents, Multiplication, Division, Addition, Subtraction.

Also, it is appropriate to reduce fractions before performing other operations, because the fraction notation is a special kind of bracketing (like parentheses).

EXAMPLE A

$\dfrac{3}{4} + \dfrac{1}{2} \div 2 - 4 =$　　A. $-3\dfrac{1}{8}$　　　　B. -3　　　C. $\dfrac{1}{4}$　　　D. $-\dfrac{5}{8}$

EXAMPLE A SOLUTION

$\dfrac{3}{4} + \dfrac{1}{2} \div 2 - 4$

Perform division before addition or subtraction.

$= \dfrac{3}{4} + \dfrac{1}{2} \div \dfrac{2}{1} - 4$

Note: $\dfrac{1}{2} \div \dfrac{2}{1} = \dfrac{1}{2} \times \dfrac{1}{2}$

$= \dfrac{3}{4} + \dfrac{1}{2} \times \dfrac{1}{2} - 4$

Note: $\dfrac{1}{2} \times \dfrac{1}{2} = \dfrac{1}{4}$

$= \dfrac{3}{4} + \dfrac{1}{4} - 4$

Perform additions and subtractions in the order that they occur from left to right.
Note: $\dfrac{3}{4} + \dfrac{1}{4} = 1$

$= 1 - 4 = -3$

The correct choice is B.

EXAMPLE B

$-3^2 - 2(3x + x) - 4 \div 2 =$

A. $1 - 8x$ B. $5 - 6x$ C. $-13 - 6x$ D. $-11 - 8x$

SOLUTION

$-3^2 - 2(3x + x) - 4 \div 2$

Simplify the expression within parentheses first.
$3x + x = 4x$

$-3^2 - 2(4x) - 4 \div 2$

Perform exponentiation before multiplication or division.
$-3^2 = -9$

$= -9 - 2(4x) - 4 \div 2$

Perform multiplications and divisions in the order in which they occur from left to right, before additions or subtractions.

$= -9 - 8x - 4 \div 2$

$= -9 - 8x - 2$

Perform subtractions in the order in which they occur from left to right.
If necessary, regroup in order to combine like terms.
$-9 - 2 = -11$

$= -11 - 8x$

This cannot be further simplified.

The correct answer is D.

For more practice with this skill go to pages 194 and 206 of this manual.

CLAST SKILL I.C.3
The student will use scientific notation in calculations involving very large or very small measurements.

Refer to Section 5.6 of *Thinking Mathematically*.

This skill will test your ability to convert numbers from decimal notation to scientific notation, and from scientific notation to decimal notation. It will also test your ability to multiply or divide numbers that have exponents, and your understanding of the associative and commutative properties of multiplication.

One idea that underlies this skill is that in some cases involving multiplication or division of certain "very large or very small" numbers, it may be convenient to first rewrite the numbers in scientific notation or another form that is closely related to scientific notation.

EXAMPLE A

$$(8.5 \times 10^6) \times (3 \times 10^{-5}) =$$

A. 2.55   B. 25.5   C. 255   D. −25.5

SOLUTION

$(8.5 \times 10^6) \times (3 \times 10^{-5})$

This computation involves just one operation: multiplication. For this reason we can use the associative and commutative properties of multiplication to regroup the factors.

$= (8.5 \times 3) \times (10^6 \times 10^{-5})$

Multiply.

$(8.5)(3) = 25.5$

$(10^6 \times 10^{-5}) = 10^{6+(-5)} = 10^1$

$= 25.5 \times 10$

$= 255$

The correct choice is C.

EXAMPLE B

$.000066 \div 11,000,000 =$

A. -6×10^{-12} B. 6×10^{-12} C. 6 D. 6×10^{12}

SOLUTION

$.000066 \div 11,000,000$

If we observe that it is easy to divide 66 by 11, we can simplify this by writing both numbers in forms that are similar to scientific notation and are related to the numbers 11 and 66.

$.000066 = 66 \times 10^{-6}$

$11,000,000 = 11 \times 10^{6}$

$= \left(66 \times 10^{-6}\right) \div \left(11 \times 10^{6}\right)$

$= \dfrac{66 \times 10^{-6}}{11 \times 10^{6}}$

$= \dfrac{66}{11} \times \dfrac{10^{-6}}{10^{6}}$ $\dfrac{b^{n}}{b^{m}} = b^{n-m}$

$= 6 \times 10^{-6-6}$

$= 6 \times 10^{-12}$ The correct choice is B.

For more practice with this skill go to page 203 of this manual.

CLAST SKILL I.C.4a
The student will solve linear equations.

Refer to Section 6.2 of *Thinking Mathematically.*

You will be given a "simple linear equation" to solve. The constants and coefficients will be integers, while the solutions may be rational numbers that aren't integers. The State of Florida specifications for this item suggest, without stating so explicitly, that these equations will not include identities or contradictions (that is, will not include equations that are true for all real numbers, nor equations that have no solutions, respectively).

The variable may occur as many as three times in the equation, and the equation may have as many as two sets of parentheses or other grouping symbols.

EXAMPLE A

If $6 - 2(x + 2) = 4(x - 3) - 2x$ then

A. $x = 0$

B. $x = \dfrac{7}{2}$

C. $x = -\dfrac{7}{2}$

D. $x = 4$

SOLUTION

$6 - 2(x + 2) = 4(x - 3) - 2x$	Clear the parentheses by using the distributive property on both sides of the equation.
$6 - 2x - 4 = 4x - 12 - 2x$	Combine like terms on both sides.
$2 - 2x = 2x - 12$	Subtract 2 from each side.
$-2x = 2x - 12 - 2$	

$-2x = 2x - 14$ Subtract $2x$ from each side.

$-2x - 2x = -14$

$-4x = -14$ Divide both sides by -4.

$\dfrac{-4x}{-4} = \dfrac{-14}{-4}$ Simplify.

$x = \dfrac{7}{2}$ The correct choice is B.

For more practice with this skill go to page 210 of this manual.

CLAST SKILL I.C.4b
The student will solve linear inequalities.

Refer to Section 6.5 of *Thinking Mathematically.*

You will be given a "simple linear inequality" to solve.

The variable may occur as many as three times in the inequality, and there may be as many as two sets of parentheses or other grouping symbols.

EXAMPLE B

If $3(4 - x) \geq 5(2x + 3)$, then

A. $x \leq \dfrac{3}{13}$

B. $x \geq \dfrac{3}{13}$

C. $x \geq -\dfrac{3}{13}$

D. $x \leq -\dfrac{3}{13}$

SKILL I.C.4

EXAMPLE B SOLUTION

$3(4-x) \geq 5(2x+3)$ 　　　　　　　Clear the parentheses by using the distributive property on both sides.

$12-3x \geq 10x+15$ 　　　　　　　Subtract 12 from each side.

$-3x \geq 10x+15-12$

$-3x \geq 10x+3$ 　　　　　　　　Subtract $10x$ from each side.

$-10x-3x \geq 3$

$-13x \geq 3$ 　　　　　　　　　　Divide both sides by -13. Remember to reverse the sense of the inequality when multiplying or dividing both sides of an inequality by a negative number.

$\dfrac{-13x}{-13} \leq \dfrac{3}{-13}$ 　　　　　　Simplify.

$x \leq -\dfrac{3}{13}$ 　　　　　　　　The correct choice is D.

For more practice with this skill go to page 215 of this manual.

CLAST SKILL I.C.5
The student will use given formulas to compute results when geometric measurements are not involved.

Refer to Sections 6.1 and 6.2 of *Thinking Mathematically*.

You will be given a problem "involving the use of a given relationship or formula." The formula will contain a dependent variable and one to three independent variables, and will be limited to first-degree and second-degree expressions involving positive rational numbers. You will be given specified values for all but one of the variables, and asked to find the value of the other variable.

EXAMPLE

Given $x = 5 + 4y - 2z$, if $x = 2$ and $z = -2$ then $y =$

A. 9   B. $-\dfrac{7}{4}$ C. $\dfrac{1}{4}$ D. 17

SOLUTION

$x = 5 + 4y - 2z$	Substitute 2 for x and -2 for z.
$2 = 5 + 4y - 2(-2)$	Simplify the right side.
$2 = 5 + 4y + 4$	
$2 = 9 + 4y$	Subtract 9 from each side.
$2 - 9 = 4y$	
$-7 = 4y$	Divide both sides by 4.
$\dfrac{-7}{4} = \dfrac{4y}{4}$	
$-\dfrac{7}{4} = y$	The correct choice is B.

For more practice with this skill go to page 207 of this manual.

CLAST SKILL I.C.6
The student will find particular values of a function.

Refer to Section 7.1 of *Thinking Mathematically*.

You will be given a function in $f(x)$ notation and a specific value to substitute for x. The function $f(x)$ will be either linear, quadratic or cubic, and will have integer coefficients. The value to be substituted for x will be a rational number from -5 to 5.

EXAMPLE

Find $f\left(-\dfrac{3}{2}\right)$, given $f(x) = 2 + 6x - 8x^2$

A. 16   B. -20 C. -25 D. 11

SOLUTION

$f(x) = 2 + 6x - 8x^2$

Substitute $-\dfrac{3}{2}$ for every occurrence of x in the formula for $f(x)$.

$2 + 6\left(-\dfrac{3}{2}\right) - 8\left(-\dfrac{3}{2}\right)^2$

Simplify the expressions involving parentheses, exponents.

$= 2 - \dfrac{18}{2} - 8\left(\dfrac{9}{4}\right)$

$= 2 - \dfrac{18}{2} - \dfrac{72}{4}$

$= 2 - 9 - 18$

$= -25$ The correct choice is C.

For more practice with this skill go to page 224 of this manual.

CLAST SKILL I.C.7
The student will factor a quadratic expression.

Refer to Section 6.6 of *Thinking Mathematically*.

You will be given a quadratic trinomial expression of the form $ax^2 + bx + c$ and asked to select an expression that is a linear factor of the given quadratic. This will require you to factor the quadratic expression.

Recall that factoring a quadratic trinomial is largely a matter of trial and error. You make a reasonable guess as to the possible factors, and then check your guess by multiplication using F.O.I.L. If your guess is incorrect, you try again. Although some quadratic expressions cannot be factored (they are called *prime* expressions), any expression that you encounter in this CLAST skill will be factorable. For these CLAST problems, the leading coefficient (the "a" in "$ax^2 + bx + c$,") will not be 1.

EXAMPLE
Which is a linear factor of the following expression? $6x^2 - 13x - 15$

A. $x - 3$ B. $2x + 5$ C. $6x + 5$ D. $x - 15$

SOLUTION

We want to factor this trinomial into the product of two linear binomials:
$(Ax + B)(Cx + D)$

Notice that in order to obtain the leading coefficient of 6, it must be the case that $AC = 6$. In order to obtain the ending coefficient of -15, it must be the case that $BD = -15$.

That means that in our trial and error process, there are only a few reasonable possibilities that need to be considered:

$(6x + 1)(x - 15)$	$(6x - 1)(x + 15)$	$(6x + 3)(x - 5)$	$(6x - 3)(x + 5)$
$(6x - 5)(x + 3)$	$(6x + 15)(x - 1)$	$(6x - 15)(x + 1)$	$(3x + 1)(2x - 15)$
$(3x + 3)(2x - 5)$	$(3x - 3)(2x + 5)$	$(3x + 5)(2x - 3)$	$(3x - 5)(2x + 3)$
$(3x - 15)(2x + 1)$	$(6x + 5)(x - 3)$	$(3x - 1)(2x + 15)$	$(3x + 15)(2x - 1)$

Each of these possible factorizations will yield the correct leading coefficient (6) and the correct ending coefficient (-15); it is a matter of checking, via F.O.I.L., which factorization will produce the correct middle coefficient (-13).

Persistent checking reveals that the correct factorization is $(6x + 5)(x - 3)$. The correct choice is A.

For more practice with this skill go to page 217 of this manual.

CLAST SKILL I.C.8
The student will find the roots of a quadratic equation.

Refer to Section 6.6 of *Thinking Mathematically*.

You will be given a quadratic equation and asked to find its real number solutions (its "real roots").

All of these problems can be solved by using the **quadratic formula**:

If $ax^2 + bx + c = 0$, then

$$x = \frac{-b \pm \sqrt{b^2 - 4ac}}{2a}$$

In some cases (but not all) is may also be possible to solve the equation by factoring.

Here are a couple of guiding facts to bear in mind for this CLAST skill:

i. The equation $ax^2 + bx + c = 0$ can be solved by factoring if and only if the number $b^2 - 4ac$ is a perfect square.

ii. Since the equations for this SKILL will have only real number solutions, the number $b^2 - 4ac$ will always be non-negative.

EXAMPLE A

Find the correct solutions to
$6x - 1 = x^2$

A. $3 + 2\sqrt{2}$ and $3 - 2\sqrt{2}$

B. $-3 + \sqrt{10}$ and $-3 - \sqrt{10}$

C. $3 + 4\sqrt{2}$ and $3 - 4\sqrt{2}$

D. $-3 + 4\sqrt{2}$ and $-3 - 4\sqrt{2}$

EXAMPLE A SOLUTION

First we put the equation in standard form by subtracting x^2 from each side.

$$-x^2 + 6x - 1 = 0$$

We will use the quadratic formula: $x = \dfrac{-b \pm \sqrt{b^2 - 4ac}}{2a}$, where $a = -1$, $b = 6$, $c = -1$.

$$\dfrac{-6 \pm \sqrt{(6)^2 - (4)(-1)(-1)}}{2(-1)}$$ Simplify.

$$\dfrac{-6 \pm \sqrt{36 - 4}}{-2}$$

$$\dfrac{-6 \pm \sqrt{32}}{-2}$$ Simplify the radical part, using the fact that $\sqrt{32} = \sqrt{16} \cdot \sqrt{2} = 4\sqrt{2}$.

$$\dfrac{-6 \pm 4\sqrt{2}}{-2}$$ Factor the numerator (−2 is a factor of both terms in the numerator).

$$\dfrac{-2\left(3 \pm 2\sqrt{2}\right)}{-2}$$ Cancel the common factor of −2 from the numerator and denominator.

$$3 \pm 2\sqrt{2}$$ Write two distinct roots.

$$3 + 2\sqrt{2} \text{ and } 3 - 2\sqrt{2}$$ The correct choice is A.

Note: the fact that $b^2 - 4ac$ is not equal to a perfect square indicates that it is not possible to solve this equation by factoring.

SKILL I.C.8

EXAMPLE B

Find the correct solutions to
$2x^2 + 8x - 3 = x + 1$

A. -1 and $\dfrac{-8+\sqrt{22}}{2}$ and $\dfrac{-8-\sqrt{22}}{2}$

B. $\dfrac{-9+\sqrt{97}}{4}$ and $\dfrac{-9-\sqrt{97}}{4}$

C. $\dfrac{-7+\sqrt{65}}{4}$ and $\dfrac{-7-\sqrt{65}}{4}$

D. -4 and $\dfrac{1}{2}$

EXAMPLE B SOLUTION

We first put the equation in standard form by subtracting $x + 1$ from each side.

$2x^2 + 7x - 4 = 0$

We will use the quadratic formula: $x = \dfrac{-b \pm \sqrt{b^2 - 4ac}}{2a}$, where $a = 2$, $b = 7$, $c = -4$.

$\dfrac{-7 \pm \sqrt{(7)^2 - (4)(2)(-4)}}{2(2)}$

Simplify.

$\dfrac{-7 \pm \sqrt{49 + 32}}{4}$

$\dfrac{-7 \pm \sqrt{81}}{4}$

Simplify the radical part, using the fact that $\sqrt{81} = 9$.

$\dfrac{-7 \pm 9}{4}$

Write two distinct roots.

$\dfrac{2}{4}$ and $-\dfrac{16}{4}$

Reduce both fractions.

$\dfrac{1}{2}$ and -4

The correct choice is D.

Note: the fact that $b^2 - 4ac$ is equal to a perfect square (81) indicates that it is possible to solve this equation by factoring.

For more practice with this skill go to page 219 of this manual.

CLAST SKILL I.C.9
The student will solve a system of two linear equations in two unknowns.

Refer to Section 7.3 of *Thinking Mathematically*.

You will be given a system of two linear equations in two unknowns. The system will be presented in this form:

$ax + by = r$
$cx + dy = s$

where the coefficients a, b, c, d, r, and s will be integers whose values may range from − 25 to 25.

Because of the form in which the system is presented, it will be best to use the addition method to find the solution.

You will be asked specifically to find the "solution set" for the system of equations. This means that the solutions will be written as sets whose elements are ordered pairs.

Recall that there are three possible solution sets for a system of two linear equations in two variables:

i. The solution set may consist of exactly one (x, y) pair. Geometrically, this represents the point in the plane where two lines intersect.

ii. The solution set may be the empty set; i.e. there is no solution. Geometrically, this means that the two equations represent distinct parallel lines. You can recognize this event if in using the addition method all of the variable terms cancel and you are left with a false statement of the form $0 = N$, where N is a nonzero number.

iii. The solution set may contain infinitely many (x, y) points. Geometrically, this means that the two equations actually represent the same line; thus, any (x, y) point on that line will be a solution to the system of equations. You can recognize this event if in using the addition method all of the terms on both sides of the equals sign cancel, and you are left with a true statement of the form $0 = 0$. In this case you can use any statement equivalent to either of the given equations to express the set of all (x, y) solutions.

SKILL I.C.9

EXAMPLE A

1. Choose the correct solution set for the system of linear equations.
$4x + 2y = -6$
$5x + 5y = 10$

A. $\{(-2, 1)\}$

B. $\left\{\left(-\dfrac{1}{2}, -2\right)\right\}$

C. $\{(-5, 7)\}$

D. The empty set

EXAMPLE A SOLUTION

$4x + 2y = -6$	We will eliminate y.
$5x + 5y = 10$	Multiply the first equation by -5 and multiply the second equation by 2.
$-20x - 10y = 30$	
$10x + 10y = 20$	Add the two equations.
$-10x \quad\quad = 50$	Solve for x.
$x = -5$	Let $x = -5$ in $4x + 2y = -6$.
$4(-5) + 2y = -6$	
$-20 + 2y = -6$	Solve for y.
$2y = 14$	
$y = 7$	The solution occurs when $x = -5$, $y = 7$.
	The correct choice is A.

EXAMPLE B

1. Choose the correct solution set for the system of linear equations.
$12x - 4y = 20$
$9x - 3y = 5$

A. $\{(2, 1)\}$ B. $\{(-9, -8)\}$

C. $\left\{(x, y) \middle| y = 3x - \dfrac{5}{3}\right\}$ D. The empty set

EXAMPLE B SOLUTION

$12x - 4y = 20$ We will eliminate y.
$9x - 3y = 5$ Multiply the first equation by -3 and multiply the second equation by 4.

$-36x + 12y = -60$
$36x - 12y = 20$ Add the two equations.

$0 = -40$ Since use of the addition method has resulted in the case where all of the variables cancel and we are left with a false statement ($0 = -40$), we know that the system of equations has no solution.

The correct choice is D.

EXAMPLE C

1. Choose the correct solution set for the system of linear equations.
$x - 5y = 3$
$-4x + 20y = -12$

A. $\left\{\left(\dfrac{1}{2}, -\dfrac{1}{2}\right)\right\}$

B. $\{(8, 1)\}$

C. $\left\{(x, y) \middle| y = \dfrac{1}{5}x - \dfrac{3}{5}\right\}$

D. The empty set

SKILL I.C.9

EXAMPLE C SOLUTION

$$\begin{aligned} x - \quad 5y &= \quad 3 \\ -4x + 20y &= -12 \end{aligned}$$

We will eliminate x.
Multiply the first equation by 4.

$$\begin{aligned} 4x - \ 20y &= \quad 12 \\ -4x + 20y &= -12 \end{aligned}$$

Add the two equations.

$$0 = 0$$

Since use of the addition method has resulted in the case where all of the terms cancel on both sides of the equals sign, leaving the true statement $0 = 0$, we know that there are infinitely many solutions.

Choice C is the only option that expresses infinitely many solutions, so the correct answer must be C. Nevertheless, we will verify that it is correct.

Choose either of the original equations and solve it for y in terms of x.

$$x - 5y = 3$$

Subtract x from both sides.

$$-5y = -x + 3$$

Divide both sides by -5.

$$y = \frac{-x + 3}{-5}$$

$$y = \frac{x}{5} - \frac{3}{5}$$

This equation gives the relationship between the x and the y coordinates of every solution point. The solution set may be expressed as

$$\left\{ (x, y) \middle| y = \frac{1}{5}x - \frac{3}{5} \right\}$$

The correct choice is C.

For more practice with this skill go to page 225 of this manual.

CLAST SKILL II.C.1
The student will use properties of operations correctly.

Refer to Sections 5.5 and 6.1 of *Thinking Mathematically.*

You will be given either algebraic expressions or equations. In some cases, you may be asked to choose the expression or equation which is algebraically correct; in other cases, you will be asked to choose the expression or equation which is not algebraically correct.

The problems will test your ability to recognize the correct use of the following properties of real numbers:

i.   Commutative property of addition or multiplication;

ii. Associative property of addition or multiplication;

iii. Identity property for addition or multiplication;

iv. Inverse property for addition or multiplication;

v.   Distributive property of multiplication over addition.

Although the problems will test your ability to recognize the correct use of these properties, you will not have to know the names of the properties.

EXAMPLE A

Choose the expression equivalent to the following:

$(4b^2)(a^3b)$

A. $4b^2 + a^3b$

B. $4b^2a^3 + 4b^3$

C. $(a^3b)(4b^2)$

D. $(4a^2)(b^3a)$

SKILL II.C.1

EXAMPLE A SOLUTION

Choice C is equivalent to the given expression, because of the commutative property of multiplication (the order in which the two factors are listed has been reversed). None of the other choices is equivalent to the given expression.

The correct choice is C.

EXAMPLE B

Select the equation that is <u>not</u> true for all real numbers.

A. $3x + 3y = 3(x + y)$

B. $4 + (x + y) = (4 + x) + y$

C. $x(1) = 1(x) = x$

D. $(3x)y = 3(x + y)$

EXAMPLE B SOLUTION

Choice A illustrates the distributive property of multiplication over addition.

Choice B illustrates the associative property of addition.

Choice C illustrates the identity property of multiplication.

Choice D is an incorrect attempt to use an associative or distributive property.

The correct choice (the equation that is <u>not</u> true for all real numbers) is D.

For more practice with this skill go to page 208 of this manual.

CLAST SKILL II.C.2
The student will determine whether a particular number is among the solutions of a given equation or inequality.

Refer to Sections 6.2, 6.5, and 6.6 of *Thinking Mathematically.*

You will be given two equations and one inequality, along with a number. By substituting the number into each of the equations and the inequality, you will determine in each case whether the number is a solution of the equation or inequality.

The number will be an integer between −10 and 10, or a fraction whose numerator and denominator are both less than 5.

The equations and inequality may be linear or quadratic, and may involve absolute value.

EXAMPLE

For each of the statements given below, determine whether $x = \dfrac{3}{2}$ is a solution.

    i. $|4 - x| < 2$
    ii. $2x + 1 = 6x - 5$
    iii. $2x^2 + x - 5 = 0$

A. i only   B. ii only   C. iii only   D. ii and iii only

SOLUTION

We will substitute $\dfrac{3}{2}$ for x in each of the three statements.

i. $|4 - x| < 2$

$\left|4 - \dfrac{3}{2}\right| < 2\,?$

$\left|\dfrac{8}{2} - \dfrac{3}{2}\right| < 2\,?$

$\left|\dfrac{5}{2}\right| < 2\,?$

$|2.5| < 2\,?$

$2.5 < 2\,?$ False

ii. $2x + 1 = 6x - 5$

$2\left(\dfrac{3}{2}\right) + 1 = 6\left(\dfrac{3}{2}\right) - 5$?

$\dfrac{6}{2} + 1 = \dfrac{18}{2} - 5$?

$3 + 1 = 9 - 5$?

$4 = 4$?   True

iii. $2x^2 + x - 5 = 0$

$2\left(\dfrac{3}{2}\right)^2 + \dfrac{3}{2} - 5 = 0$?

$2\left(\dfrac{9}{4}\right) + \dfrac{3}{2} - 5 = 0$?

$\dfrac{18}{4} + \dfrac{3}{2} - 5 = 0$?

$\dfrac{18}{4} + \dfrac{6}{4} - \dfrac{20}{4} = 0$?

$\dfrac{18 + 6 - 20}{4} = 0$?

$\dfrac{4}{4} = 0$?

$1 = 0$?   False

The correct choice is B.

For more practice with this skill go to page 221 of this manual.

CLAST SKILL II.C.3
The student will recognize statements of proportionality and variation.

Refer to Section 6.4 of *Thinking Mathematically.*

You will be given a situation involving conditions of direct or inverse variation and asked to identify the response that mathematically represents the conditions. This is not a skill that actually requires that you solve a word problem. Instead, it is a skill that presents a word problem and tests your ability to recognize the algebraic expressions that would be used to solve the problem.

The problem may contain irrelevant information.

EXAMPLE A

The value of a certain bond is directly proportional to the number of years since the bond was purchased. After four years the bond is worth $2620, and the bond-owner also owns two cars. Let V be the value of the bond after nine years. Select the correct statement of the given conditions.

A. $\dfrac{V}{4} = \dfrac{2620}{9}$

B. $V = \dfrac{2620}{36}$

C. $\dfrac{V}{2620} = \dfrac{18}{4}$

D. $\dfrac{V}{9} = \dfrac{2620}{4}$

EXAMPLE A SOLUTION

We can set up a proportion by using the fact that the value is directly proportional to the number of years:

$$\frac{\text{value of bond after 9 nine years}}{9 \text{ years}} = \frac{\text{value of bond after 4 nine years}}{4 \text{ years}}$$

We can substitute specific numbers for three of the four quantities in the proportion, and substitute V for the other quantity.

SKILL II.C.3

$$\frac{V}{9} = \frac{2620}{4}$$

The correct choice is D.

We must realize that any equation that is algebraically equivalent to this proportion is also a correct answer.

Using the cross products principle, we see that the correct answer may also have been written as

$$4V = (9)(2620)$$

Dividing both sides of this equation by 2620V, we see that another correct answer could have been

$$\frac{4}{2620} = \frac{9}{V}.$$

There are several other correct possibilities for the answer.

Also note that we weren't asked to solve for V. We were just asked to show that we are capable of recognizing a correct way to algebraically "set up" the solution to the problem.

Finally, that fact that the bond-owner also owns two cars was irrelevant.

EXAMPLE B

The number of days that it takes for a team of workers to build a house varies inversely as the number of workers on the team. It takes 18 days for a team of 6 workers to build a house if they are working at the job 5 days per week. Let D be the number of days it takes for a team of 10 workers to build a similar house working 5 days per week. Select the correct statement of the given condition.

A. $\dfrac{D}{10} = \dfrac{18}{6}$

B. $\dfrac{6}{D} = \dfrac{10}{18}$

C. $\dfrac{D}{5} = \dfrac{10}{18}$

D. $\dfrac{30}{6} = \dfrac{D}{18}$

EXAMPLE B SOLUTION

The significant fact here is that the number of days required to build the house varies inversely as the number of workers on the team (the fact that they are working 5 days per week is irrelevant).

Remember the general rule for proportions in which "*y* varies inversely as *x*."

$$\frac{\text{The first value for } y}{\substack{\text{The value for } x \text{ corresponding} \\ \text{to the second value for } y}} = \frac{\text{The second value for } y}{\substack{\text{The value for } x \text{ corresponding} \\ \text{to the first value for } y}}$$

In this case, the role of *y* is played by "the number of days" and the role of *x* is played by "the number of workers," so one correct proportion would be:

$$\frac{18}{10} = \frac{D}{6}$$

This isn't one of the choices listed, however.

It is important to recognize that any proportion that is algebraically equivalent to this would also be correct. For example, if we apply the cross-products principle, we have this equivalent statement:

$$(6)(18) = 10D$$

If we divide both sides of this equation by 18D we have this proportion:

$$\frac{6}{D} = \frac{10}{18}$$

The correct choice is B.

For more practice with this skill go to page 213 of this manual.

CLAST SKILL II.C.4
The student will identify regions of the coordinate plane which correspond to specified conditions and vice-versa.

Refer to Section 7.4 of *Thinking Mathematically*.

This skill tests your ability to graph systems of linear inequalities. It also tests your knowledge of linear equations and their graphs.

The problem might give you a system of linear inequalities and ask you to select the correct graph for the system, or it might show you the graph of a shaded region of the plane and ask you to select the corresponding system of linear inequalities.

The system may involve one, two or three inequalities.

One important note: in *Thinking Mathematically*, the solution to a system of two or more inequalities is obtained by finding the intersection of the various shaded regions. On the CLAST we find the intersection of two or more shaded regions when the inequalities are separated by the word "AND;" on the other hand, if the inequalities are separated by the word "OR," then we find the union of two or more shaded regions.

EXAMPLE A

Identify the conditions which correspond to the shaded region of the plane.

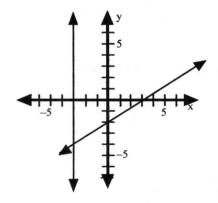

A. $x \geq -3$ and $2x - 3y \leq 6$

B. $x \geq -3$ and $3x - 2y \leq 6$

C. $y \geq -3$ and $2x - 3y \leq 6$

D. $y \geq -3$ and $3x - 2y \leq 6$

EXAMPLE A SOLUTION

We start by identifying the equations for the boundary lines of the shaded region.

The leftmost boundary line is a vertical line whose x-intercept is -3.

Recall that any vertical line will have an equation of the form $x = k$, where k is the x-intercept. This means that the equation for this boundary line is $x = -3$.

Since the boundary line is solid rather than dashed, the corresponding inequality must be either $x \geq -3$ or $x \leq -3$.

If we use the coordinates of the point $(0, 0)$ for test purposes, we see that the correct inequality is $x \geq -3$.

Notice that we have eliminated choices C and D.

To identify the second inequality, we find the equation for the second boundary line. We can first determine its slope: notice that as we go from the y-intercept to the x-intercept we have a *rise* of 2 units and a *run* of 3 units.

Thus we have the following slope for the second boundary line:

$$m = \frac{rise}{run} = \frac{2}{3}$$

We see that the second boundary line has a slope of 2/3 and a y-intercept of -2.

According to the slope-intercept formula, the equation for this line is
$$y = \frac{2}{3}x - 2$$

We will put this equation in standard form, $Ax + By = C$.

Multiply both sides by 3:

$3y = 2x - 6$

Subtract $2x$ from both sides:

$-2x + 3y = -6$

Multiply both sides by -1:

$2x - 3y = 6$.

SKILL II.C.4

Since the equation for this boundary line is $2x - 3y = 6$ and the boundary line is solid rather than dashed, the corresponding inequality is either $2x - 3y \leq 6$ or $2x - 3y \geq 6$.

(Notice that this eliminates choice B.)

By using the coordinates of the point $(0, 0)$ for test purposes, you can verify that the correct inequality is $2x - 3y \leq 6$. Thus the system of inequalities is

$x \geq -3$ and $2x - 3y \leq 6$.

The correct choice is A.

EXAMPLE B

Which shaded region identifies the portion of the plane which corresponds to the conditions $x + y \leq -2$ and $x - y \leq 2$?

A.

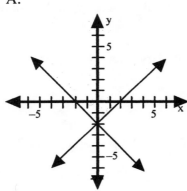

B.

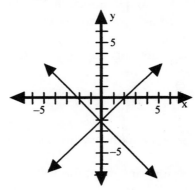

C.

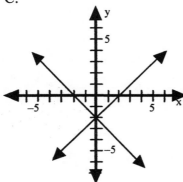

D.

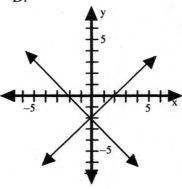

92

EXAMPLE B SOLUTION

We will first identify the shaded region for $x + y \leq -2$.

The boundary line is the line $x + y = -2$.

We will graph this line by finding its intercepts.

To find the x-intercept we let $y = 0$ and solve for x:

$x + 0 = -2$

The x-intercept is -2.

To find the y-intercept we let $x = 0$ and solve for y:

$0 + y = -2$

The y-intercept is -2.

Now we graph the line whose x-intercept is -2 and y-intercept is -2:

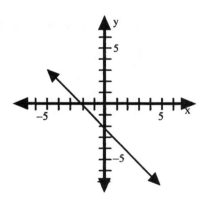

We use a solid line rather than a dashed line because the inequality symbol is "$\leq$" as opposed to "$<$."

The inequality $x + y \leq -2$ will consist of the points on this line along with either all of the points above the line or all of the points below the line. To decide which points to shade we can use the coordinates of the point $(0, 0)$ for test purposes.

Letting $x = 0$ and $y = 0$ in $x + y \leq -2$ we have

$0 + 0 \leq -2?$

$0 \leq -2?$

This statement is false. Since the inequality is false for the point (0, 0), it will also be false for all of the other points that are above the boundary line. This means that it will be true for all of the points below the boundary line:

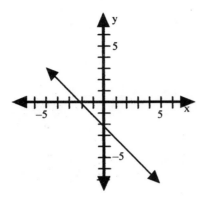

Next we identify the shaded region for $x - y \leq 2$.

We start by graphing the boundary line, $x - y = 2$.

Following the same procedure as outlined above, we will find that the x-intercept is 2 and the y-intercept is -2.

We graph the line having those intercepts:

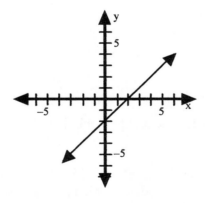

The inequality $x - y \leq 2$ will include all of the points on this line along with either all of the points above the line or all of the points below the line. Using the coordinates of the point (0, 0) to test the inequality we have:

$0 + 0 \leq 2?$

This is true. Since the inequality is true for the point $(0, 0)$, it will also be true for all of the other points that are above the boundary line:

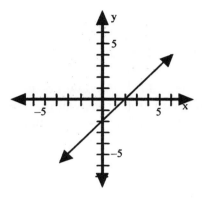

The shaded region for the system of inequalities $x + y \leq -2$ and $x - y \leq 2$ will be the intersection of the two shaded regions shown above:

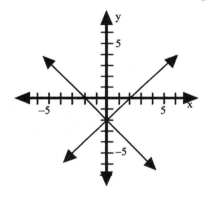

The correct choice is C.

Note: if the system of inequalities had been expressed as $x + y \leq -2$ **or** $x - y \leq 2$, then the solution region would be the union of the two solution regions shown above.

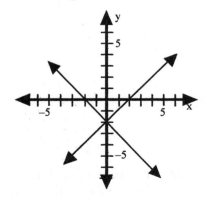

For more practice with this skill go to page 228 of this manual.

CLAST SKILL III.C.2
The student uses applicable properties to select equivalent equations and inequalities.

Refer to Sections 6.2 and 6.5 of *Thinking Mathematically.*

This skill tests your ability to simplify equations and inequalities. You will be given an equation or inequality and asked to select an equivalent equation or inequality. The equivalent equation or inequality will be one that would result from correctly performing one algebraic operation on the given expression.

The following algebraic properties may be involved:

a. $a = b$ if and only if $a + c = b + c$;
b. $a = b$ if and only if $ac = bc$, $c \neq 0$;
c. $a > b$ if and only if $a + c > b + c$;
d. $a > b$ if and only if $ac > bc$, and $c > 0$;
e. $a > b$ if and only if $ac < bc$, and $c < 0$;
f. If $a > b$ and $b > c$, then $a > c$;
g. If $a = b$ and $b = c$ then $a = c$;
h. $a = b$ if and only if $b = a$;

The equations and inequalities may be linear or quadratic.

EXAMPLE A

Choose the equation equivalent to the following:

$5x - 7 = 4 - 9x$

A. $x - 7 = -9x$

B. $x - 7 = 4$

C. $x - 7 = 4 - \dfrac{9}{5}x$

D. $x - \dfrac{7}{5} = \dfrac{1}{5}(4 - 9x)$

EXAMPLE A SOLUTION

We will examine each of the multiple-choice options.

Choice A, $x - 7 = -9x$, is incorrect, because we have subtracted $4x$ from the left side, but we have subtracted 4 from the right side. We must add or subtract exactly the same quantity from each side.

Choice B, $x - 7 = 4$, is incorrect, because we have subtracted $4x$ from the left side but we have subtracted $(-9x)$ from the right side.

Choice C, $x - 7 = 4 - \dfrac{9}{5}x$, is incorrect because we have multiplied one term on each side by 1/5. If we are going to multiply both sides of an equation by some factor, we must multiply every term by that factor.

Choice D, $x - \dfrac{7}{5} = \dfrac{1}{5}(4 - 9x)$, is correct because we have multiplied both terms on the left side by 1/5, and we are also multiplying both terms on the right side by 1/5.

The correct choice is D.

EXAMPLE B

Choose the inequality equivalent to the following:

$x + 3 < 2 - 4x$

A. $3 > 2 - 5x$ B. $3 < 2 - 3x$

C. $-x - 3 > -2 + 4x$ D. $-x - 3 < -2 + 4x$

EXAMPLE B SOLUTION

We will examine each of the multiple-choice options.

Choice A, $x + 3 < 2 - 4x$, is incorrect because we have subtracted x from both sides but we have also reversed the sense of the inequality. We should reverse the sense on the inequality only in the case where we have multiplied both sides by a negative quantity.

Choice B, $3 < 2 - 3x$, is incorrect because we have subtracted x from the left side, but we have added x to the right side.

Choice C, $-x - 3 > -2 + 4x$, is correct because we have multiplied both sides by -1 and reversed the sense of the inequality.

Choice D, $-x - 3 < -2 + 4x$, is incorrect because we have multiplied both sides by -1 but neglected to reverse the sense of the inequality.

The correct choice is C.

For more practice with this skill go to pages 209 and 215 of this manual.

CLAST SKILL IV.C.1

The student solves real-world problems involving the use of variables, aside from commonly-used geometric formulas.

Refer to Sections 6.2, 6.3, and 6.4 of *Thinking Mathematically*.

You will be given a real-world problem taken from business, social studies, industry, education, economics, environmental studies, the arts, physical science, sports, or a consumer-related context. You will not need to have any special knowledge of the area in order to solve the problem.

The mathematical structure of the problem will involve quadratic or linear relationships, proportions, or variation.

The numerical content will involve rational numbers.

The problem may contain irrelevant information.

EXAMPLE A

James is making pizzas for a party. Each pizza requires $3\frac{1}{2}$ cups of flour for dough, $\frac{1}{2}$ lb. of mozzarella cheese, 20 oz. of tomato sauce, 2 tbs. of olive oil, 1 cup of mushrooms, 6 oz. of pepperoni, and serves four adults or six children. To make these pizzas he is using a total of $17\frac{1}{2}$ cups of flour. How many people are expected for the party, assuming that they are all adults?

A. 20   B. 5   C. 25   D. 14

EXAMPLE A SOLUTION

Let x be the number of pizzas that he will make.

Since each pizza serves 4 adults, $4x$ is the number of adults expected.

Since each pizza requires $3\frac{1}{2}$ cups of flour, we can find x by solving the following equation:

$$3\frac{1}{2}x = 17\frac{1}{2}$$

It will be easier to solve this equation if we write the mixed fractions as decimal numbers:

$3.5x = 17.5$

$$x = \frac{17.5}{3.5} = 5$$

He is making 5 pizzas, so the number of adults expected is

$(5)(4) = 20$

The correct choice is A.

(Notice that much of the information in the problem was irrelevant.)

EXAMPLE B

When the local chapter of a political discussion group elects their officers, they choose one executive, five representatives, and two senators. The national convention for this organization elects officers in the same proportions as the local chapters. How many senators does the national convention have if they have 120 representatives?

A. 24 B. 240 C. 300 D. 48

EXAMPLE B SOLUTION

Let x be the number of senators in the national convention.

The number of senators is proportional to the number of executives and it is also proportional to the number of representatives. Since we are told that there are 115 representatives in the national convention, we can find the number of senators by using this proportion:

$$\frac{\text{number of senators in national convention}}{\text{number of representatives in national convention}} = \frac{\text{number of senators in local chapter}}{\text{number of representatives in local chapter}}$$

Using our variable and the specific numbers given in the problem, we have this equation:

$$\frac{x}{115} = \frac{2}{5}$$

We solve for x by multiplying both sides of the equation by 115:

$$x = \frac{2(115)}{5}$$

$$x = \frac{230}{5}$$

$$x = 48$$

The correct choice is D.

EXAMPLE C

The formula $S = \frac{1}{2}gt^2$ gives the distance an object falls from rest during t seconds. In this formula S is measured in feet and g, the gravitational constant, is 32 ft/sec². How many seconds will it take for an object to fall 576 feet?

A. 36 seconds

B. 6 seconds

C. 294,912 seconds

D. 24 seconds

EXAMPLE C SOLUTION

We will substitute 576 for S and 32 for g in the formula $S = \frac{1}{2}gt^2$, and then solve for t.

$576 = \frac{1}{2}(32)t^2$	Simplify the right side: $\frac{1}{2}(32) = 16$
$576 = 16t^2$	Divide both sides by 16.
$\frac{576}{16} = t^2$	Simplify the left side: $576/16 = 36$
$36 = t^2$	Subtract 36 from each side.
$0 = t^2 - 36$	Factor.
	Set each factor equal to zero.
$0 = (t + 6)(t - 6)$	
$t + 6 = 0 \qquad t = -6$	It does not make sense for the solution to
$t - 6 = 0 \qquad t = 6$	be a negative number, so the correct choice is B.

For more practice with this skill go to pages 212, 214, and 223 of this manual.

CLAST SKILL IV.C.2
The student solves problems that involve the logic and structure of algebra.

Refer to Section 6.3 of *Thinking Mathematically*.

You will be given a word problem "involving words, symbols or relations" and asked to select the symbolic expression or statement that satisfies the requirements of the problem. These will not be "real-world" word problems, but rather more abstract problems about algebra and properties of numbers. You will not be asked to solve the problem, either, but will be asked instead to choose an algebraic expression that summarizes the conditions described in the problem.

EXAMPLE A

Choose the equation that is equivalent to the verbal description:

When a number n is reduced by 4, the result is the same as the product of 3 and the quantity obtained when the number is doubled and increased by 1.

A. $4 - n = 3 + 2(n + 1)$

B. $n - 4 = 3(2n + 1)$

C. $4 - 3n = 2n + 1$

D. $\dfrac{4}{n} = 3 + (2n + 1)$

EXAMPLE A SOLUTION

We will translate the verbal description into an algebraic expression.

"a number n is reduced by 4"	$n - 4$
"the result is the same as"	$=$
"the product of 3 and"	$3 \times$
"the quantity obtained when the number is doubled and increased by 1"	$2n + 1$

Rewriting this as one expression, without the "times" sign, we have

$n - 4 = 3(2n + 1)$

The correct choice is C.

(Notice that we are not asked to solve this equation.)

SKILL IV.C.2

EXAMPLE B

A two digit positive number is equal to 4 times the sum of its digits. Which equation should be used to find its digits, x and y?

A. $xy = 4x + y$ 　　　 B. $xy = 4(x + y)$ 　　 C. $10x + y = 4(x + y)$ 　　　 D. $x + y = 4xy$

EXAMPLE B SOLUTION

If x is the tens digit and y is the ones digit, then the two-digit number is equal to

$10x + y$ 　　　　(for instance, 24 is equal to $2(10) + 4$; 85 is equal to $8(10) + 5$).

If the number is equal to four times the sum of its digits, then

$10x + y = 4(x + y)$

The correct choice is C.

EXAMPLE C

Choose the equation that is equivalent to the verbal description:

For any positive integer n, if the integer is increased by one and squared the result is always greater than the result when the number is squared and increased by one.

A. $n^2 + 1 > 2n^2$ 　　　 B. $(n + 1)^2 > 2n$ 　　 C. $n^2 + 1 > (n + 1)^2$ 　 D. $(n + 1)^2 > n^2 + 1$

EXAMPLE C SOLUTION

We will translate the verbal description into an algebraic expression.

"...the integer is increased by one and squared" 　　　　　　　　　$(n + 1)^2$

"the result is always greater than" 　　　　　　　　　　　　　　　$>$

"the result when the number is squared and increased by one" 　　　$n^2 + 1$

Combining these into a single expression we have

$(n + 1)^2 > n^2 + 1$ 　　　　　　　The correct choice is D.

Again, note that we are not asked to solve anything, but simply to translate stated conditions into algebraic expressions.

For more practice with this skill go to page 211 of this manual.

CLAST SKILL I.D.1
The student will identify information contained in bar, line and circle graphs.

Refer to Sections 1.2, 8.1 and 12.1 of *Thinking Mathematically.*

You will be given a labeled bar, line or circle graph. Based on the information in the graph you will either be asked to determine a specific frequency or percentage or, given a frequency or percentage you will be asked to identify a specific category from the graph. These problems will require that you perform no more than two computations, such as finding the sum or difference of frequencies, determining a percentage from a frequency, or determining a frequency from a percentage.

EXAMPLE A

The line graph below tracks the attendance for a math class during the days of a week during summer semester. How many more students attended on Wednesday than attended of Monday?

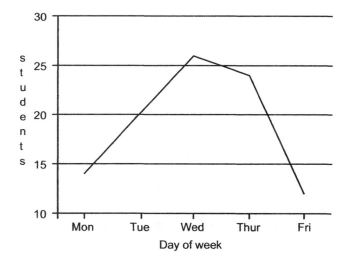

A. 26   B. 12   C. 3   D. 6

EXAMPLE A SOLUTION

The graph shows that approximately 26 students attended on Wednesday and approximately 14 students attended on Monday.

$26 - 14 = 12$ The best choice is B.

SKILL I.D.1

EXAMPLE B

The circle graph below shows preferences of a number of fast food consumers.
What percent prefer cheeseburgers?

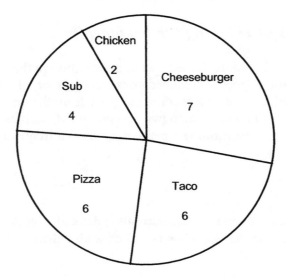

A. 7%

B. 70%

C. 28%

D. 72%

EXAMPLE B SOLUTION

In order to determine the percentage, we first need to find n, the number of people surveyed.

$n = 4 + 2 + 7 + 6 + 6 = 25$

There were 25 people surveyed.

Among these 25 people, 7 of them prefer cheeseburgers.

$7/25 = .28 = 28\%$

The correct choice is C.

EXAMPLE C

The bar graph below shows the distribution according to age of the children in a summer camp. For what age is it true that 14 campers are less than that age?

A. 10

B. 12

C. 6

D. 7

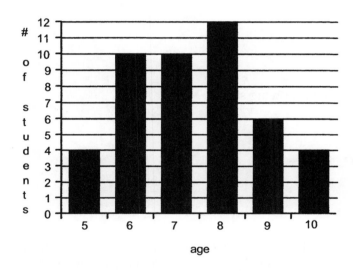

EXAMPLE C SOLUTION

The graph shows that 4 children are 5 years old and 10 children are 6 years old. This means that 14 children are less than 7 years old.

The correct choice is D.

For more practice with this skill go to pages 239 and 322 of this manual.

CLAST SKILL I.D.2
The student will determine the mean, median and mode of a set of numbers.

Refer to Section 12.2 of *Thinking Mathematically.*

You will be given a collection of whole numbers. The list will contain at least eight but at most eleven numbers. The numbers may range in value from one to thirty. You will be asked to select the mean, median or mode for the collection of numbers.

For these CLAST problems there will be a unique mode, and the mean, median and mode will all have different values.

EXAMPLE A

What is the <u>median</u> of the data in the following sample?

5, 8, 1, 8, 4, 3, 6, 7

A. 6   B. 5.25   C. 8   D. 5.5

EXAMPLE A SOLUTION

The median is determined by the number or numbers in the middle of the list when the numbers are put in numerical order. We begin by putting the numbers in numerical order:

1, 3, 4, 5, 6, 7, 8, 8

There are eight numbers (data points) on the list. Since there are an even number of data points, the median will be the average of the two middle data points:

1, 3, 4, **5, 6,** 7, 8, 8

The two middle numbers are 5 and 6, so the median is $(5 + 6)/2 = 11/2 = 5.5$.

The correct choice is D.

For further practice verify that the mean is 5.25 and the mode is 8.

EXAMPLE B

What is the <u>mean</u> of the data in the following sample?

10, 16, 12, 10, 18, 12, 10, 10, 18, 14

A. 14   B. 13   C. 15   D. 10

EXAMPLE B SOLUTION

There are ten data points on this list, so the mean is the sum of all data points, divided by 10.

Mean = (10 + 16 + 12 + 10 + 18 + 12 + 10 + 10 + 18 + 14)/10

= 130/10

= 13

The correct choice is B.

For further practice verify that the median is 12 and the mode is 10.

EXAMPLE C

What is the <u>mode</u> of the data in the following sample?

30, 22, 26, 23, 30, 22, 22, 22, 28

A. 22   B. 25   C. 23   D. 30

EXAMPLE C SOLUTION

In this list of nine numbers the most frequently occurring number is 22, so the mode is 22.

The correct choice is A.

For further practice verify that the mean is 25 and the median is 23.

For more practice with this skill go to page 325 of this manual.

CLAST SKILL I.D.3
The student will use the fundamental counting principle.

Refer to Sections 11.1, 11.2 and 11.3 of *Thinking Mathematically*.

Despite what the title may suggest, this skill could involve the fundamental counting principle, permutations or combinations.

You will be given a word problem referring to a process in which a subset is being chosen from a given set, or in which a series of decisions are described. You will be asked to determine the number of possible outcomes. The problem will not involve gambling scenarios or games of chance.

The numbers in these problems will be rather small. For processes in which a subset is being selected from a given set, the given set will contain at most six elements. For problems involving a series of decisions, there will be at most four decisions with no more than five options for each decision.

EXAMPLE A

The Locale County School Board consists of four members. Tonight they will vote on a proposal to hire a new contractor to provide school lunches. Each member may vote "yea," "nay" or "abstain." How many outcomes are possible?

A. 12

B. 7

C. 81

D. 64

EXAMPLE A SOLUTION

Since each of the four members must vote, this is a process involving four decisions. For each of the four decisions there are three different options. According to the fundamental counting principle the number of outcomes is

$(3)(3)(3)(3) = 81$.

The correct choice is C.

EXAMPLE B

There are six members of the board of directors of a charitable organization. From among their group they will select four officers: Chairperson, Associate Chairperson, Secretary and Treasurer. How many outcomes are possible, assuming that no person will hold more than one position?

A. 360

B. 1296

C. 24

D. 18

EXAMPLE B SOLUTION

Since the situation involves choosing four people from a set of six people and arranging the four people according to who receives which title, the number of outcomes is the number of permutations of six things taken four at a time.

$$_6P_4 = \frac{6!}{(6-4)!} = \frac{6!}{2!} = \frac{720}{2} = 360$$

The correct choice is A.

Note: problems that can be solved with the permutation formula can also be solved with the fundamental counting principle. In this case, we need to make four dependent decisions:

i. Choose Chairperson: 6 options.

ii. Choose Associate Chairperson: 5 options.

iii. Choose Secretary: 4 options.

iv. Choose Treasurer: 3 options.

According to the fundamental counting principle, the number of outcomes is

$(6)(5)(4)(3) = 360$.

SKILL I.D.3

EXAMPLE C

An office employs six women and three men. Two women and two men will be selected to attend an employee-relations seminar. In how many ways may the selection be made?

A. 180

B. 72

C. 45

D. 90

EXAMPLE C SOLUTION

Choosing two women involves choosing two people from a set of six people. Since they are both receiving the same treatment (attend a seminar) the order in which they are chosen or listed is not important. Therefore, the number of ways to choose two women is the number of combinations of six things taken two at a time.

$$_6C_2 = \frac{6!}{(6-2)!2!} = \frac{6!}{4!\,2!} = \frac{720}{24 \cdot 2} = \frac{720}{48} = 15$$

Likewise, the number of ways to choose two men from the set of three men is the number of combinations of three things taken two at a time.

$$_3C_2 = \frac{3!}{(3-2)!2!} = \frac{3!}{1!\,2!} = \frac{6}{1 \cdot 2} = \frac{6}{2} = 3$$

According to the fundamental counting principle the number of ways to choose two women and two men is

$(15)(3) = 45.$

The correct choice is C.

For more practice with this skill go to pages 295-300 of this manual.

CLAST SKILL II.D.1
The student will recognize properties and interrelationships among the mean, median and mode in a variety of distributions.

Refer to Section 12.2 of *Thinking Mathematically.*

You will be given information about a distribution of numbers and asked to identify a statement that correctly states a relationship between two of the measures of central tendency. The information about the data distribution may be strictly verbal or it may refer to a graph. Specific frequencies will not be provided, so you will not necessarily be able to compute the mean and there will not be sufficient information to compute the median via the usual method.

The distribution will be either symmetric ("approximately normal"), skewed to the left, or skewed to the right. Some pertinent facts are outlined below.

When a distribution is symmetric with respect to the mode, we say that it is "approximately normal," and in such a case the numbers "balance out" so that the mean, median and mode are all the same.

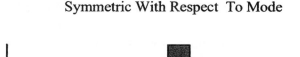

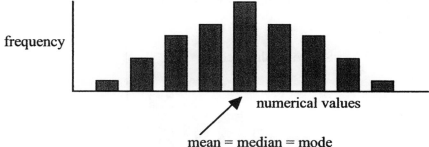

SKILL II.D.1

In a case where the mode is the *smallest* number in the distribution and as values increase their frequencies decrease, we say that the distribution is "skewed to the right." In such a distribution the mean must be greater than the mode (since the mode is the smallest number and the average number must be greater than the smallest number in the distribution), and likewise the median is either greater than or equal to the mode.

Skewed to the Right

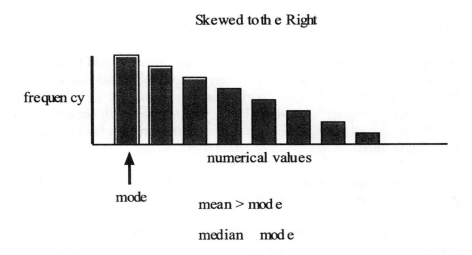

If the mode is the greatest number in the distribution and as values increase their frequencies increase, we say that the distribution is "skewed to the left." In such a case the mean must be less than the mode (since the mode is the largest number and the average number must be less than the largest number) and likewise the median is less than or equal to the mode.

Skewed to the Left

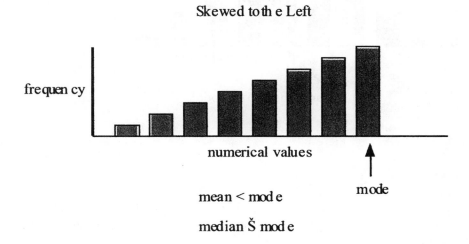

112

EXAMPLE A

Incoming freshmen at Greatbig University must take a mathematics placement test. The students are classified into groups according to the results of the test. (Group 1 represents the lowest scores and Group 5 represents the highest scores.) The distribution is shown in the graph below.

Which of the following statements is true about this distribution?

A. The median is the same as the mean.

B. The median is the same as the mode.

C. The mean is greater than the mode.

D. The median is less than the mode.

EXAMPLE A SOLUTION

In one of these problems it will always be easy to identify the mode. If the problem presents graphical information, the mode must be the value under the tallest column. In this case the mode is 1. Note that 1 is also the smallest number in the distribution. In any collection of numbers that are not all identical, the average number (the mean) must be greater than the smallest number (which happens to be the mode in this example), so it is correct to observe "the mean is greater than the mode" which is choice C.

The correct choice is C.

Notice that this is an example of data skewed to the right. According to the earlier discussion we know that the following statement would also be true:

"The median is greater than the mode." (In the special case where the column above the mode is taller than the heights of all of the other columns combined, it would be correct to say, "the median is the *same as* the mode").

Moreover, usually (but not necessarily always) in a case where data is skewed to the right, the median will be less than the mean (that is, the median tends to fall between the mode and the mean).

SKILL II.D.1

EXAMPLE B

Almost half of the participants in a youth league all-star baseball tournament are 10 years old. A lesser number are 9 years old and fewer still are 8 years old. Which of the following statements is true about this distribution?

A. The mean is greater than the mode.

B. The median is the same as the mean.

C. The median is the same as the mode.

D. The mode is greater than the median.

EXAMPLE B SOLUTION

The description indicates that there are more 10-year-olds than there are 9-year-olds or 8-year-olds, so the mode is 10. Note that 10 is also the largest number in the distribution. It must be correct to say that "the mean is less than the mode" since in any collection of numbers that are not all the same the average number must be smaller than the greatest number in the collection. This eliminates choice A.

It is also correct to observe that in a collection of numbers that are not all identical the median must be less than (or equal to) the largest number in the distribution. In this case it is reasonable to say, "the median is less than the mode" which is the same as saying "the mode is greater than the median."

The correct choice is D.

Notice that if we graph the distribution we will see that it is skewed to the left, and the observations we have made agree with the general information about skewed data that were discussed earlier.

In a case like this it would be correct to say, "the mode is *the same* as the median" only if more than half of the participants were 10 years old. More generally, when graphed data is skewed to the left the mode will be the same as the median if the height of the column above the mode is greater than the combined heights of all of the other columns.

Also, usually (but not necessarily always) when a distribution is skewed to the left the mean is less than the median (that is, the median tends to lie between the mean and the mode).

For more practice with this skill go to page 329 of this manual.

CLAST SKILL II.D.2
The student will choose the most appropriate procedure for selecting an unbiased sample from a target population.

Refer to Section 12.1 of *Thinking Mathematically*.

You will be given a situation where a sample is to be selected from a target population, and asked to select the method that would be most appropriate for obtaining an unbiased sample. The target population will be clearly identified. The "most appropriate method for obtaining an unbiased sample" will be a method that does **both** of the following:

1. Selects **only** members of the target population; and

2. Makes the selection **randomly** from within the entire target population.

It will be relatively easy to choose the most appropriate method by eliminating the inappropriate choices. Any choice that does not describe a random method is eliminated, as is any choice that does not address the indicated target population.

EXAMPLE A

The publisher of a news magazine wants to survey its subscribers in order to determine their interest in an on-line version of the magazine. Which of the following procedures would be most appropriate for obtaining a statistically unbiased sample?

A. Randomly select 1000 people from a list of clients of a national Internet service provider.

B. Select 1000 people whose names are chosen from the middle of an alphabetical list of the magazine's subscribers.

C. Randomly select 1000 people from a list of all those have sent e-mails to the magazine.

D. Randomly select 1000 people from an alphabetical listing of all of the magazine's subscribers.

SKILL II.D.2

EXAMPLE A SOLUTION

Choice A does not address the target population (the magazine's subscribers).

Choice B is not sufficiently random because the selections are confined to a certain portion of the target population.

Choice C does not address the target population.

Among the given choices the most appropriate choice is D.

EXAMPLE B

A newspaper wants to find out which of its features writers are popular with its readers and decides to conduct a survey. Which of the following would be most appropriate for obtaining a statistically unbiased sample?

A. Surveying a random sample of people from the telephone directory.

B. Surveying the first one hundred subscribers from an alphabetical list of all subscribers.

C. Having people voluntarily mail in their preferences.

D. Surveying a random sample of readers from a list of all subscribers.

EXAMPLE B SOLUTION

Choice A does not address the target population.

Choice B is not sufficiently random.

Choice C does not address the target population and is not random.

Among the given choices, the most appropriate method is choice D.

For more practice with this skill go to page 316 of this manual.

CLAST SKILL II.D.3
The student will identify the probability of a specified outcome in an experiment.

Refer to Sections 11.4, 11.5, 11.6 and 11.7 of *Thinking Mathematically*.

You will be given a description of a random process and asked to identify the probability of a specified event. The random process will not involve cards, dice or other gambling scenarios. You may need to use counting techniques to determine the number of outcomes. The problems will not involve odds.

You will need to know the following formulas:

$P(E) = n(E)/n(S)$

$P(\text{not } E) = 1 - P(E)$

$P(E \text{ or } F) = P(E) + P(F) - P(E \text{ and } F)$

$P(E \text{ or } F) = P(E) + P(F)$ if E and F are mutually exclusive.

$P(E \text{ and } F) = P(E) \times P(F, \text{ given } E)$

$P(E \text{ and } F) = P(E) \times P(F)$ if E and F are independent

EXAMPLE A

There are eight puppies and six kittens at the animal shelter. If one animal is randomly selected, what is the probability that it is a kitten?

A. 4/3   B. 3/4   C. 4/7   D. 3/7

EXAMPLE A SOLUTION

Since there are 14 animals from which to choose, when one animal is randomly chosen there are 14 possible (equally likely) outcomes. Of these 14 animals, six are kittens. Therefore,

$P(\text{kitten}) = 6/14 = 3/7$

The correct choice is D.

SKILL II.D.3

EXAMPLE B

There are four Democrats and three Republicans on the City Council. A developer is going to select three of these people and give them campaign donations. If the selection is random, what is the probability that all three recipients will be Democrats?

A. 4/7   B. 3/7   C. 1/7   D. 4/35

EXAMPLE B SOLUTION

There are seven people to choose from, so if three people are randomly chosen the number of outcomes is

$$_7C_3 = \frac{7!}{(7-3)!\,3!} = \frac{7!}{4!3!} = \frac{7 \cdot 6 \cdot 5 \cdot 4!}{4!3!} = \frac{7 \cdot 6 \cdot 5}{3!}$$

$$= \frac{7 \cdot 6 \cdot 5}{6} = 35$$

There are 35 possible outcomes in this experiment.

Since four of the members are Democrats, the number of ways to choose three Democrats is

$$_4C_3 = \frac{4!}{(4-3)!\,3!} = \frac{4!}{1!3!} = \frac{24}{6} = 4$$

Thus,

P(three Democrats are chosen) = 4/35

The correct choice is D.

EXAMPLE C

In a group of 28 students who are taking their math test today, four forgot to bring pencils, two forgot to bring a calculator, and one forgot both pencils and a calculator. If one student is selected, find the probability that he or she forgot to bring pencils or a calculator.

A. 1/4   B. 3/14   C. 5/58   D. 3/28

118

EXAMPLE C SOLUTION

Let E be the event "forgot pencils."

Let F be the event "forgot calculator."

Then according to the information above

$P(E) = 4/28$

$P(F) = 2/28$

$P(E \text{ and } F) = 1/28$

So,

$P(E \text{ or } F) = P(E) + P(F) - P(E \text{ and } F) = 4/28 + 2/28 - 1/28$

$= (4 + 2 - 1)/28$

$= 5/28$

The correct choice is C.

Alternative solution

When a counting problem or probability problem refers to two categories that overlap ("forgot pencils" and "forgot calculator" in this case) we can use a Venn diagram to organize the information.

We start by identifying the number that goes in the innermost region.

On e forgot both p encils and c alculator.

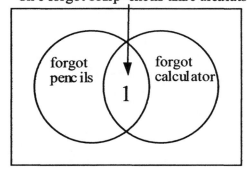

Next, use the fact that four forgot to bring pencils.

Four forgot pencils.

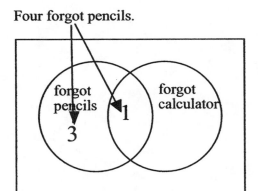

Now we use the fact that two forgot to bring a calculator.

Two forgot a calculator.

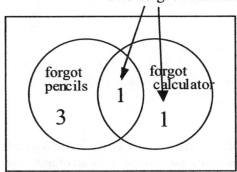

Next, use the fact that the diagram represents a total of 28 students in order to find the number that goes in the region of the diagram exterior to the two circles.

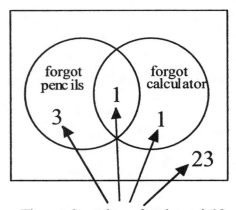

The total number of students is 28.

Now we use the diagram to identify the number of students who satisfy the compound condition "forgot pencils or forgot calculator."

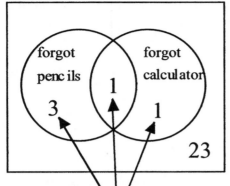

These five are the ones who forgot pencils or forgot a calculator.

The diagram shows that among the 28 students there were five who forgot pencils or forgot a calculator. Thus,

P(forgot pencils or forgot calculator) = 5/28.

The correct choice is C.

EXAMPLE D

In a group of 28 students who are taking their math test today, four forgot to bring pencils, two forgot to bring a calculator, and one forgot both pencils and a calculator. If one student is selected, find the probability that he or she forgot to bring pencils but didn't forget to bring a calculator.

A. 1/3

B. 3/28

C. 1/28

D. 26/28

SKILL II.D.3

EXAMPLE D SOLUTION

To solve this problem we want to construct a Venn diagram. Refer to the Venn diagram constructed in the solution for EXAMPLE C and use it to identify the portion of the population satisfying the compound condition "forgot pencils but didn't forget calculator."

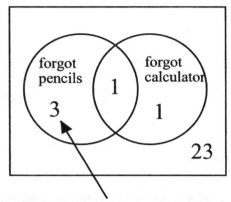

These three are the ones who forgot
pencils but didn't forget a calculator.

The diagram shows that out of a population of 28 students there were 3 who forgot pencils but didn't forget a calculator. Thus,

P(forgot pencils but didn't forget calculator) = 3/28.

The correct choice is B.

EXAMPLE E

There are three women and five men who are eligible to be selected Secretary and Treasurer of the Chamber of Commerce. If the two positions are filled by random drawing, and nobody can hold more than one position, find the probability that both the Secretary and Treasurer will be women.

A. 1/4   B. 3/8   C. 2/5   D. 3/28

EXAMPLE E SOLUTION

Consider this to be a two-stage process. Stage one involves choosing the Secretary and stage two involves choosing the Treasurer.

Let E be the event the Secretary is a woman. Since there are eight people to choose from and three are women,

$P(E) = 3/8$

Let F be the event that the Treasurer is a woman. Assuming that the person selected to be Secretary was a woman, then there are seven people left and two of them are women, so

$P(F, \text{ given } E) = 2/7$

Finally,

$P(E \text{ and } F) = P(E) \times P(F, \text{ given } E) = \dfrac{3}{8} \times \dfrac{2}{7} = \dfrac{6}{56} = \dfrac{3}{28}$

The correct choice is D.

EXAMPLE F

According to estimates from the Centers for Disease Control, among the men who contracted HIV in 1998, 27% were injecting drug users. If we randomly select two men who contracted HIV in 1998, what is the probability that both were injecting drug users? *(Source: Health, United States, 1998, Centers for Disease Control, National Center for HIV, STD and TB Prevention, Division of HIV/AIDS.)*

A. .54   B. .27   C. .0729   D. .73

EXAMPLE F SOLUTION

On the CLAST, *when a probability problem refers to randomly selecting two or more individuals from a (large) population of unspecified size, the intention is that we treat the outcomes as independent events.* It is worth mentioning that this assumption is not entirely mathematically correct, although as a practical matter it will not usually result in a significant degree of error. Having pointed this out, we proceed.

In the experiment consisting of independently choosing two people from the population of men who contracted HIV in 1998, let E be the event that the first man is an injecting drug user and let F be the event that the second man is an injecting drug user. According to the population statistics cited above,

$P(E) = .27$ and $P(F) = .27$, so

$P(E \text{ and } F) = P(E) \times P(F) = .27 \times .27 = .0729$

The correct choice is C.

For more practice with this skill go to pages 304-305 and 309-312 of this manual.

CLAST SKILL III.D.1
The student infers relations and makes accurate predictions from studying statistical data.

Refer to Sections 12.1 and 12.5 of *Thinking Mathematically*.

These problems are similar to those in Skill I.D.1 in that you will be shown "real-world" statistical data presented in graphical form. The problems in this skill tend to emphasize qualitative rather than quantitative observations. In this skill, for instance, you may be asked to identify or use a trend in the data, whereas the problems in Skill I.D.1 involve more straightforward computations of frequencies or percentages.

EXAMPLE A

The graph below shows information about the racial breakdown of people living below the officially established poverty level in the U.S. *(Source: Bureau of the Census, U.S. Department of Commerce.)*

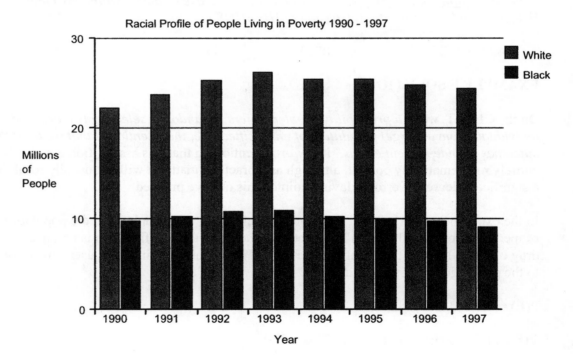

Select the statement that correctly describes a trend in the data.

A. Whenever poverty among whites increases poverty among blacks decreases.

B.   Poverty among whites has been increasing in the most recent years depicted.

C. In every year depicted there have been more blacks than whites living in poverty.

D. Poverty among blacks has been decreasing in the most recent years depicted.

EXAMPLE A SOLUTION

Among the four given statements the only correct statement is choice D.

Note that the problem involves terminology that is somewhat imprecise or informal, such as "the most recent years." This is indicative of the general spirit of this CLAST skill.

EXAMPLE B

The scatter plot below relates the number of violent crimes reported to the number of property crimes reported by year over the years 1977 – 1996. *(Source: FBI Uniform Crime Reports 1996.)*

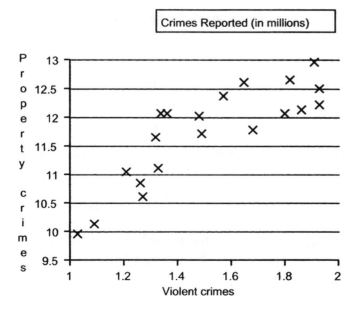

SKILL III.D.1

Which of the following best describes a relationship between property crimes and violent crimes?

A. Increasing numbers of property crimes cause increasing numbers of violent crimes.

B. Increasing numbers of violent crimes cause decreasing numbers of property crimes.

C. There appears to be a positive association between numbers of violent crimes and numbers of property crimes.

D. There appears to be a negative association between numbers of violent crimes and numbers of property crimes.

EXAMPLE B SOLUTION

In a problem like this any statement that infers a cause-and-effect relationship between the two variables (such as we see in choices A and B) will be incorrect. Although there may be an apparent trend in the data, that trend alone will never be sufficient to support the inference that a change in one variable is *causing* (as opposed to *being associated with*) a change in the other variable.

A correct observation would be "An increase in property crime is associated with an increase in violent crime."

Another correct observation would be "An increase in violent crime is correlated with an increase in property crime."

In general, when a scatter diagram shows a pattern that tends to rise as we look from left to right, we say that there is a *positive correlation or association* between the two variables.

When a scatter diagram shows a pattern that tends to fall as we look from left to right we say that there is a *negative correlation or association* between the two variables.

The correct choice is C.

EXAMPLE C

Referring to the data in EXAMPLE B above, what is the least number of property crimes reported in a year where the number of violent crimes was at least 1.6 million?

A. 11.8 million

B. 12.6 million

C. 12.4 million

D. 10 million

EXAMPLE C SOLUTION

First, on the scatter plot identify all of the cases that satisfy the given condition "the number of violent crimes was at least 1.6 million." Those would be any occurrences that are on or to the right of the vertical line representing 1.6 million violent crimes.

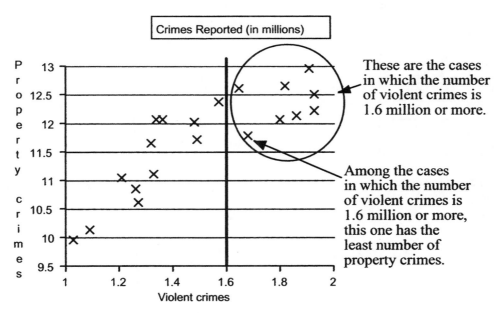

We see that among those cases in which the number of violent crimes was 1.6 million or more, the least number of property crimes is roughly 11.8 million.

The best choice is A.

For more practice with this skill go to pages 318 and 337 of this manual.

CLAST SKILL IV.D.1
The student interprets real-world data involving frequency and cumulative frequency tables.

Refer to Sections 12.2 and 12.4 of *Thinking Mathematically.*

You will be presented with numerical "real-world data" in the form of a **relative** frequency table, a **cumulative** frequency table, or a table showing **percentile** rank. Depending upon the style of data table, you may be asked to find the mean, median or mode, or you may be asked to identify the percentage of the population that falls into a specified interval.

A **relative** frequency table differs from a frequency table in that it describes a distribution in terms of proportions or percents rather than giving specific frequencies.

EXAMPLE A

Workers who are being trained to perform a certain manufacturing job are given a dexterity test. The scores on this test vary from 10 (lowest) to 15 (highest). The distribution of scores is shown in the following table.

score	proportion
10	.28
11	.24
12	.18
13	.12
14	.10
15	.08

Find the mean score.

A. 12.5

B. 11.76

C. 12

D. 10

EXAMPLE A SOLUTION

Note that finding the mean for data in a relative frequency table differs slightly from finding the mean of data in a frequency table, because we do not know the size of the population (this would be the sum of the frequencies in a frequency table). In a relative frequency table like this, the sum of the proportions should always be 1 (or, if the proportions are given as percents, the sum of the percents should be 100%).

To find the mean we find the sum of all terms formed by multiplying a value by its proportion.

Mean = 10(.28) + 11(.24) + 12(.18) + 13(.12) + 14(.10) + 15(.08)

= 2.80 + 2.64 + 2.16 + 1.56 + 1.40 + 1.20

= 11.76

The correct choice is B.

EXAMPLE B

Referring to the data in EXAMPLE A, find the median.

A. 12

B. 10

C. 11

D. 12.5

EXAMPLE B SOLUTION

The median is determined by the value in the "middle of the list" when the values are ranked in numerical order. Note that in this table the values (test scores) are already in numerical order, from least (10, at the top of the table) to greatest (15, at the bottom of the table). Frequency and relative frequency tables are almost always set up so that the values are in numerical order.

We need to locate the "middle" of the distribution (that is, we need to find the value such that 50% of the distribution is at or below that value and 50% is at or above that value). We can do this by adding to the table a column for **cumulative proportion** and using it to observe the value at which cumulative proportion meets or exceeds .50 (or 50%).

score	proportion	cumulative proportion
10	.28	.28
11	.24	.52
12	.18	.70
13	.12	.82
14	.10	.92
15	.08	1.00

Notice how the column for cumulative proportion was constructed:

1. In the first row, cumulative proportion is the same as proportion.

2. In every other row, cumulative proportion is equal to that row's proportion plus the previous row's cumulative proportion.

score	proportion	cumulative proportion
10	.28	.28
11	.24	*.52*
12	.18	.70
13	.12	.82
14	.10	.92
15	.08	1.00

Cumulative proportion first exceeds .50 (50%) in the second row, where the corresponding test score is 11. Thus, the median is 11.

The correct choice is C.

Note: in a problem like this if cumulative proportion was exactly equal to .50 (exactly equal to 50%) in some row, the median would be the average of the value from that row and the value from the next row.

Also, for the data in this table the mode is 10 (the mode is always the value having the greatest proportion or percent).

EXAMPLE C

Referring to the data in EXAMPLE A above, find the test score such that 82% of the workers have scores less than or equal to that score.

A. 15   B. 14   C. 12   D. 13

EXAMPLE C SOLUTION

Referring to the column for cumulative proportion in the solution to EXAMPLE B, we see that a score of 13 has a cumulative proportion of .82. This means that 82% of the test-takers had scores of 13 or lower.

EXAMPLE D

An ice-cream vendor sells ice-cream cones whose prices are $1.00, $1.50, $2.00 and $2.50. The table below shows the distribution of products sold, according to price.

price ($)	percent
1.00	20%
1.50	25%
2.00	20%
2.50	35%

Find the median price ($).

A. 1.50   B. 2.00   C. 1.75   D. 2.50

EXAMPLE D SOLUTION

The values (prices) are already given in numerical order, so to find the median we will add to the table a column for cumulative percent and use it to identify the value at which cumulative percent first meets or exceeds 50%.

price ($)	percent	cumulative percent
1.00	20%	20%
1.50	25%	45%
2.00	20%	*65%*
2.50	35%	100%

Cumulative percent first meets or exceeds 50% in the row of the table where the value (price) is 2.00, so the median is 2.00.

The correct choice is B.

SKILL IV.D.1

EXAMPLE E

Referring to the data in EXAMPLE D above, find the mean.

A. 1.75   B. 2.50   C. 1.85   D. .25

EXAMPLE E SOLUTION

In order to use the technique described in EXAMPLE A above, we will rewrite the percents as equivalent decimal numbers (proportions).

price ($)	proportion
1.00	.20
1.50	.25
2.00	.20
2.50	.35

To find the mean, we compute the sum of all terms formed by multiplying a value (price) by its proportion.

Mean = (1.00)(.20) + (1.50)(.25) + (2.00)(.20) + (2.50)(.35)

= .200 + .375 + .400 + .875 = 1.85

The correct choice is C.

For further practice verify that the mode is 2.50.

EXAMPLE F

The table below shows the percentile distribution of students according to their scores on an achievement test.

Test Score	Percentile Rank
800	99
750	90
700	85
600	75
500	50
400	25
300	15

What percentage of students had scores less than 400?

A. 25   B. 15   C. 40   D. 10

EXAMPLE F SOLUTION

Problems involving percentile rank refer to this fundamental definition:

The percentile rank of a score tells the percentage of the population that lies below the given score.

The table shows that a test score of 400 has a percentile rank of 25. This tells us directly that 25% of the test-takers had scores less than 400. The correct choice is A.

EXAMPLE G

Referring to the data in EXAMPLE F above, what percentage of the test-takers had scores greater than or equal to 750?

A. 99

B. 90

C. 9

D. 10

EXAMPLE G SOLUTION

The table shows that a test score of 750 has a percentile rank of 90. According to the definition of percentile rank, this means that 90% of test-takers had scores *less than* 750. If 90% had scores less than 750, then the other 10% (100% − 90% = 10%) had scores greater than or equal to 750.

The correct choice is D.

Note that in solving this problem we used the definition of percentile rank and the fact that "greater than or equal to" is the opposite of "less than."

SKILL IV.D.1

EXAMPLE H

Referring to the data in EXAMPLE F, what percentage of test-takers had scores between 500 and 700?

A. 75

B. 135

C. 35

D. 50

EXAMPLE H SOLUTION

The table shows that a score of 500 has a percentile rank of 50 and a score of 700 has a percentile rank of 85. According to the definition of percentile rank, 50% of test-takers had scores less than 500 and 85% of test takers had scores less than 700.

Thus, approximately

$$85\% - 50\% = 35\%$$

had scores between 500 and 700.

The best choice is C.

Important note: this problem has been written in accordance with the State of Florida's CLAST specifications, even though it is technically incorrect. It would be correct to say that 35% of test-takers had scores less than 700 but greater than *or equal to* 500. Be aware of the possibility that on the CLAST you may encounter incorrect phraseology on this type of problem.

For more practice with this skill go to pages 326 and 334 of this manual.

CLAST SKILL IV.D.2
The student solves real-world problems involving probability.

Refer to Sections 11.4, 11.5, 11.6 and 11.7 of *Thinking Mathematically.*

You will be given real-world data presented in a graph, chart or table and asked a question about probability. The problem will not refer to cards, dice or gambling scenarios.

You will need to know the following formulas:

P(E) = n(E)/n(S)

P(not E) = 1 – P(E)

P(E or F) = P(E) + P(F) – P(E and F)

P(E or F) = P(E) + P(F) *if E and F are mutually exclusive.*

P(E and F) = P(E) × P(F, given E)

P(E and F) = P(E) × P(F) *if E and F are independent.*

P(E, given F) = P(E and F)/P(F)

The problem may also ask for the expected number of occurrences of an event.

SKILL IV.D.2

EXAMPLE A

The pie chart below shows the racial/ethnic profile of homeless people in Florida *(source: Florida Department of Children and Families)*.

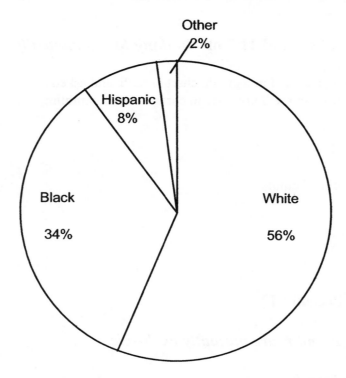

If two homeless people are randomly selected, find the probability that both of them are Hispanic.

A. .16   B. .0064   C. .64   D. .08

EXAMPLE A SOLUTION

On the CLAST, when a probability problem involves selecting two individuals from a (large) population of unspecified size, the exam writer's intention is that we treat the outcomes as independent events.

Let E be the event that the first person selected is Hispanic, and let F be the event that the second person selected is Hispanic.

Since 8% of Florida's homeless population are Hispanic,

P(E) = .08 and P(F) = .08

We are trying to find P(E and F).

$P(E \text{ and } F) = P(E) \times P(F) = .08 \times .08 = .0064$

The correct choice is B.

EXAMPLE B

Referring to the data in Example A above, if two homeless people are randomly selected, find the probability that at least one of them is black.

A. .5644   B. .68   C. .1156   D. .34

EXAMPLE B SOLUTION

Let E be the event that the first person selected is black. Since 34% of the homeless population is black, $P(E) = .34$.

Likewise, we let F be the event that the second person selected is black, and observe that $P(F) = .34$.

We are trying to find $P(E \text{ or } F)$.

$P(E \text{ or } F) = P(E) + P(F) - P(E \text{ and } F)$

$= P(E) + P(F) - P(E) \times P(F)$

$= .34 + .34 - .34 \times .34 = .68 - .1156 = .5644$

The correct choice is A.

Alternative solution

With E and F defined as above, note that $P(\text{not } E) = 1 - P(E) = 1 - .34 = .66$.

Likewise, $P(\text{not } F) = .66$.

Recall that "E or F" is the opposite (complement) of "not E and not F." Thus,

$P(E \text{ or } F) = 1 - P(\text{not } E \text{ and not } F)$

$= 1 - P(\text{not } E) \times P(\text{not } F) = 1 - (.66 \times .66) = 1 - .4356 = .5644$

Again, the correct choice is A.

SKILL IV.D.2

EXAMPLE C

Incoming business students at a community college are given a mathematics placement test. The students are classified into groups and placed into courses according to their placement test scores. Group 5 is highest and Group 1 is lowest. The results are summarized in the table below.

Group	Percent of students	Placement
1	38%	MAT1033
2	32%	MAC1105
3	15%	MAC1105
4	10%	MAC2233
5	5%	MAC2233

If 1200 students take the test, how many would we expect to score in Group 5?

A. 5　　　　B. 50　　　　C. 60　　　　D. 600

EXAMPLE C SOLUTION

The table shows that 5% are expected to finish in Group 5.

Five percent of $1200 = (.05)(1200) = 60$.

The correct choice is C.

EXAMPLE D

Referring to the table in EXAMPLE C above, if one incoming business student is randomly selected find the probability that he or she did not score in Group 5.

A. .5　　　　B. .95　　　　C. .05　　　　D. .005

EXAMPLE D SOLUTION

The table shows that 5% of incoming business students score in Group 5. This means that 95% don't score in Group 5. Thus,

P(not in Group 5) = .95.

The correct choice is B.

EXAMPLE E

Referring to the table in EXAMPLE C above, if one student is randomly selected, find the probability that he or she placed into MAC1105, given that he or she scored higher than Group 2.

A. 3/20

B. 15/15

C. 15/47

D. 1/2

EXAMPLE E SOLUTION

Let E be the event that a randomly selected student placed into MAC1105.

Let F be the event that a randomly selected student scored higher than Group 2.

We are trying to find P(E, given F).

Now, the table shows that 15% + 10% + 5% = 30% scored higher than Group 2, so

$P(F) = .30$

Also, 15% simultaneously placed into MAC1105 and had scores higher than Group 2, so

$P(E \text{ and } F) = .15$

Thus,

$$P(E, \text{ given } F) = \frac{P(E \text{ and } F)}{P(F)}$$

$$= \frac{.15}{.30} = \frac{15}{30}$$

$$= 1/2$$

The correct choice is D.

For more practice with this skill see pages 301, 306, and 312 of this manual.

CLAST SKILL I.E.1
The student will deduce facts about set-inclusion or non-set-inclusion from a diagram.

Refer to Sections 2.2 and 2.3 of *Thinking Mathematically.*

You will be given a Venn diagram containing up to four sets (including the universal set), and asked to choose a statement that correctly describes a relationship between sets in the diagram.

EXAMPLE

Sets S, T, U and V are related as shown in the diagram.

Which of the following statements is true, assuming that none of the regions of the diagram is empty?

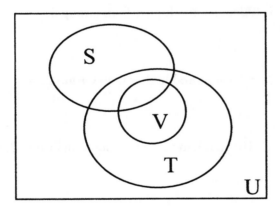

A. Any element that is a member of V is also a member of S and T.

B. There is no element that is a member of all three sets S, T and V.

C. Any element that is not a member of T is not a member of V.

D. Any element that is a member of U is a member of S or T or V.

SOLUTION

We will examine each of the multiple-choice responses.

A. Any element that is a member of V is also a member of S and T.

In order for this statement to be true, set V must be entirely contained within the intersection of sets S and T; choice A is false.

B. There is no element that is a member of all three sets S, T and V.

In order for this statement to be true, there must be no region in which all three sets overlap; choice B is false.

C. Any element that is not a member of T is not a member of V.

In order for this statement to be true, set V must be contained entirely within set T; choice C is true.

D. Any element that is a member of U is a member of S or T or V.

Since we are told that "no region of the diagram is empty," there must be some element that is within set U but outside of sets S and T and V; choice D is false.

The correct choice is C.

For more practice with this skill go to page 171 of this manual.

CLAST SKILL II.E.1
The student will identify statements equivalent to the negations of simple and compound statements.

Refer to Sections 3.1 and 3.5 of *Thinking Mathematically.*

You will be given a simple or compound statement expressed in words, and will be asked to select its negation. The statement will be a conjunction, disjunction, conditional, or quantified statement.

These problems may refer to any of the following rules of negation:

$\sim (\sim p) \equiv p$

$\sim (p \wedge q) \equiv \sim p \vee \sim q$

$\sim (p \vee q) \equiv \sim p \wedge \sim q$

$\sim (p \rightarrow q) \equiv p \wedge \sim q$

$\sim (\text{All are } p) \equiv \text{Some are } (\sim p)$

$\sim (\text{Some are } p) \equiv \text{None are } p$

$\sim (\text{Some are } p) \equiv \text{All are } (\sim p)$

EXAMPLE A

Select the statement that is the negation of the statement "Some winter days are cool."

A. All winter days are cool.

B. No winter days are cool.

C. Some winter days are not cool.

D. If it is a winter day, then it is cool.

SOLUTION

Let p represent the condition "Winter days are cool."

We are asked to find the negation of a statement having the form "Some are p."

There are two correct forms of this negation: "All are $(\sim p)$" or "None are p."

These correspond to these two statements, respectively:

"All winter days are <u>un</u>cool."

"No winter days are cool."

We see that choice B is correct.

EXAMPLE B

Select the statement that is the negation of the statement "Carrots are vegetables and gems are minerals."

A. Carrots aren't vegetables and gems aren't minerals.

B. Carrots aren't vegetables and gems are minerals.

C. Gems are minerals and carrots are vegetables.

D. Carrots aren't vegetables or gems aren't minerals.

EXAMPLE B SOLUTION

Let p be the statement "Carrots are vegetables."

Let q be the statement "Gems are minerals."

The statement "Carrots are vegetables and gems are minerals" corresponds to $p \wedge q$.

To correctly write its negation, we recall one of DeMorgan's Laws:

$$\sim (p \wedge q) \equiv \sim p \vee \sim q$$

Applying this rule to the given statement, we have this negation: "Carrots aren't vegetables or gems aren't minerals."

The correct choice is D.

EXAMPLE C

Select the statement that is the negation of the statement "If the dam is built, then the forest will be flooded."

A. If the dam is not built, then the forest will not be flooded.

B. The dam is built or the forest will not be flooded.

C. The dam is built and the forest will not be flooded.

D. If the dam is built, then the forest will not be flooded.

SKILL II.E.1

EXAMPLE C SOLUTION

Let p be the statement "The dam is built."

Let q be the statement "The forest will be flooded."

The statement "If the dam is built, then the forest will be flooded" corresponds to $p \rightarrow q$.

To correctly write its negation, we recall the rule for writing the negation of a conditional statement:

$$\sim (p \rightarrow q) \equiv p \wedge \sim q$$

Applying this rule to the given statement, we have this negation: "The dam is built and the forest will not be flooded."

The correct choice is C.

For more practice with this skill go to pages 174 and 176 of this manual.

CLAST SKILL II.E.2
The student will determine equivalence or non-equivalence of statements.

Refer to Section 3.5 of *Thinking Mathematically.*

You will be given a simple or compound statement expressed in words, and will be asked to select an equivalent (or non-equivalent) statement. The statement will be a conjunction, disjunction, or conditional statement.

These problems may refer to any of the following rules of logical equivalency.

$p \to q \equiv\, \sim p \vee q$

$p \to q \equiv\, \sim q \to\, \sim p$

$\sim (\sim p) \equiv p$

$\sim (p \wedge q) \equiv\, \sim p \vee \sim q$

$\sim (p \vee q) \equiv\, \sim p \wedge \sim q$

$\sim (p \to q) \equiv p \wedge \sim q$

Notice that many of these rules are the same rules for negations that are tested under Skill II.E.1.

EXAMPLE A

Select the statement that is logically equivalent to "If I'm expecting visitors then I wash the dishes."

A. If I wash the dishes, then I'm expecting visitors.

B. I wash the dishes, or I'm expecting visitors.

C. If I don't wash the dishes, then I'm not expecting visitors.

D. If I'm not expecting visitors, then I don't wash the dishes.

SOLUTION

Let *p* represent the statement "I'm expecting visitors."

Let q represent the statement "I wash the dishes."

The statement "If I'm expecting visitors then I wash the dishes" is symbolized $p \to q$.

SKILL II.E.2

From the facts outlined above, we see that there are two rules of logical equivalency that could be used.

1. $p \rightarrow q \equiv\, \sim p \vee q$

Applying this rule, we have the following equivalent statement: "I'm not expecting visitors or I wash the dishes." This is not among the listed choices.

2. $p \rightarrow q \equiv\, \sim q \rightarrow\, \sim p$

Applying this rule, we have the following equivalent statement: "If I don't wash the dishes, then I'm not expecting visitors." This is choice C.

The correct choice is C.

EXAMPLE B

Select the statement that is logically equivalent to "It is not true that either Fluffy or Whiskers are puppies."

A. Fluffy is a puppy and Whiskers is a puppy.

B. Fluffy is not a puppy and Whiskers is not a puppy.

C. Fluffy is not a puppy or Whiskers is not a puppy.

D. If Fluffy is not a puppy, then Whiskers is not a puppy.

EXAMPLE B SOLUTION

Let p be the statement "Fluffy is a puppy."

Let q be the statement "Whiskers is a puppy."

The statement "It is not true that either Fluffy or Whiskers are puppies" is the negation of $p \vee q$, so we are asked to find the statement that is equivalent to the *negation* of $p \vee q$.

According to one of DeMorgan's Laws, $\sim (p \vee q) \equiv\, \sim p \wedge\, \sim q$.

In words, the statement $\sim p \wedge\, \sim q$ is "Fluffy is not a puppy and Whiskers is not a puppy."

The correct choice is B.

EXAMPLE C

Select the statement that is not logically equivalent to "If you own a car, then you need insurance."

A. If you don't own a car, then you don't need insurance.

B. You don't own a car or you need insurance.

C. If you don't need insurance, then you don't own a car.

D. You need insurance or you don't own a car.

EXAMPLE C SOLUTION

Let p be the statement "You own a car."

Let q be the statement "You need insurance."

The statement "If you own a car, then you need insurance" is symbolized as $p \rightarrow q$.

Since we are asked to choose the statement that is *not* equivalent to the given statements, three of the four answers must be equivalent to $p \rightarrow q$.

Using the rules of equivalency listed above, we have the following:

1. $p \rightarrow q \equiv \sim p \vee q$ Applying this rule to the given statement, we have this equivalent statement: "You don't own a car or you need insurance." This is choice B.

2. $p \rightarrow q \equiv \sim q \rightarrow \sim p$ Applying this rule to the given statement, we have the following equivalent statement: "If I don't need insurance, then you don't own a car." This is choice C.

We need to find a third statement that is equivalent to the given conditional statement, but we have already exhausted both of the rules for equivalency for conditional statements. However, because the disjunction is symmetric, we can take choice B, "You don't own a car or you need insurance," and reverse its two components to get this equivalent statement: "You need insurance or you don't own a car." This is choice D.

We were asked to find the statement that was *not* equivalent to the given statement. We have determined that choices B, C and D *are* equivalent to the given statement, so the only remaining choice is A.

The correct choice (the non-equivalent statement) is A.

For more practice with this skill go to page 177 of this manual.

CLAST SKILL II.E.3
The student will draw logical conclusions from data.

Refer to Sections 3.6 and 3.7 of *Thinking Mathematically.*

This skill may involve analysis of arguments or some other situation in which you are supposed to make a logical conclusion based on specified conditions, rules and other information.

EXAMPLE A

Given that:

i. All crooners are singers;
ii. Eddie is not a singer;

determine which conclusion can be logically deduced.

A. Eddie is a crooner.   B. Eddie is not a crooner.
C. Andy is my crony.   D. None of the above.

EXAMPLE A SOLUTION

We will use Euler diagrams to analyze the premises. We are trying to find a valid conclusion that follows from the two premises. If there is a valid conclusion from the two premises, every Euler diagram that agrees with both premises will automatically agree with that valid conclusion. On the other hand, for any invalid conclusion it will be possible to draw a diagram that agrees with both premises yet denies that conclusion.

We begin with a diagram that agrees with the first premise.

i. All crooners are singers.

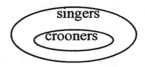

Now we expand the diagram so that it agrees with the second premise as well.

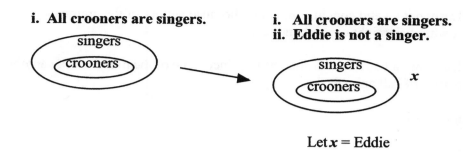

i. **All crooners are singers.**

i. **All crooners are singers.**
ii. **Eddie is not a singer.**

Let x = Eddie

Notice that any diagram that agrees with both premises will automatically agree with the conclusion "Eddie is not a crooner."

The correct choice is B.

EXAMPLE B

Given that:

i. If this animal is an otter, then it eats fish;
ii. This animal eats fish;

determine which statement can be logically deduced.

A. This animal is an otter.   B. All animals are otters.
C. Some fish eat otters.   D. None of the above.

EXAMPLE B SOLUTION

Let p be the statement "This animal is an otter."
Let q be the statement "This animal eats fish."

The premise arrangement has this symbolic form:

$p \rightarrow q$

q

We should recognize that this is the premise arrangement for **Fallacy of the Converse**, which is a form of invalid reasoning. This tells us that choice A is not a valid conclusion, because the argument

$p \rightarrow q$

$\underline{q\ \ \ \ }$ is invalid.

$\therefore p$

SKILL II.E.3

Choice C is clearly not a valid conclusion, either, because neither of the premises say anything about what fish might or might not eat.

Choice B is clearly not a valid conclusion, either, since none of the premises say anything about the properties of "all animals."

The correct choice is D.

EXAMPLE C

Read the qualifications for adoption of a dog from the animal shelter, and then decide which applicant is eligible to adopt his chosen dog.

In order to adopt a dog, one must:

i. *Have a fenced yard or a kennel measuring at least 20 feet long and 20 feet wide; and*
ii. *Be willing to have the animal spayed if it is female or neutered if it is male; and*
iii. *Agree not to keep the dog tethered or chained.*

Chester has a fenced yard but he doesn't have a kennel, and he will never keep a dog tethered or chained. He wants to adopt a female bulldog for breeding purposes.

Lester doesn't have a fenced yard, but he does have a kennel that is 15 feet long and 25 feet wide. He wants to adopt a male mixed breed dog. He agrees to have the dog neutered and agrees not to keep the dog tethered or chained.

Nestor doesn't have a fenced yard, but he does have a kennel that measures 25 feet long and 30 feet wide. He wants to adopt a female collie mix that is already spayed. He agrees not to keep the dog tethered or chained.

A. Chester   B. Lester   C. Nestor.   D. No one is eligible.

EXAMPLE C SOLUTION

Chester is not eligible because he is apparently not willing to have the female dog spayed, since he intends to breed it.
Lester is not eligible because he does not have a fenced yard and his kennel is not sufficiently long.
Nestor doesn't have a fenced yard, but he does have an acceptable kennel; since the female dog is already spayed, he is obviously willing to adopt a spayed dog; and he agrees not to keep the dog tethered or chained. He is eligible.

The correct choice is C.

For more practice with this skill go to pages 180 and 188 of this manual.

CLAST SKILL II.E.4
The student will recognize that an argument may not be valid even though its conclusion is true.

Refer to Sections 3.6 and 3.7 of *Thinking Mathematically*.

You will be given one invalid argument and three valid arguments, expressed in words. For each argument, the conclusion will be a statement that seems true or reasonable according to our everyday experience. In the case of the invalid argument, however, the true conclusion will not be logically justified according to the given premises.

The valid arguments will involve forms that can be readily analyzed with **Euler diagrams**, or will involve the forms of **Direct Reasoning**, **Contrapositive Reasoning**, or **Transitive Reasoning**.

Direct Reasoning	Contrapositive Reasoning	Transitive Reasoning
$p \rightarrow q$	$p \rightarrow q$	$p \rightarrow q$
p	$\sim q$	$q \rightarrow r$
$\therefore q$	$\therefore \sim p$	$\therefore p \rightarrow r$

It will also be useful to recognize these common patterns of **invalid** reasoning:

Fallacy of the Converse	Fallacy of the Inverse	Misuse of Transitive Reasoning	Misuse of Transitive Reasoning
$p \rightarrow q$	$p \rightarrow q$		
q	$\sim p$	$p \rightarrow q$	$p \rightarrow q$
$\therefore p$	$\therefore \sim q$	$p \rightarrow r$	$r \rightarrow q$
		$\therefore q \rightarrow r$	$\therefore p \rightarrow r$

EXAMPLE A

All of the arguments A - D have true conclusions, but one of the arguments is not valid. Select the argument that is **not** valid.

A. All potatoes are tubers and some potatoes are edible. Therefore, some tubers are edible.
B. All vipers are snakes and some reptiles aren't snakes. Therefore, some reptiles aren't vipers.
C. Some motorcycles are noisy vehicles and some motorcycles are fast vehicles. Therefore, some noisy vehicles are fast vehicles.
D. All coastal communities are flood-prone and Carrabelle is a coastal community. Therefore, Carrabelle is flood-prone.

SKILL II.E.4

EXAMPLE A SOLUTION

We will use Euler diagrams to analyze each argument.

A. All potatoes are tubers and some potatoes are edible. Therefore, some tubers are edible.

First we need to point out than when a premise is a conjunction, we can split it into two separate premises. In this case, the argument is equivalent to:

All potatoes are tubers. Some potatoes are edible. Therefore, some tubers are edible.

We will try to draw a diagram that agrees with both premises but denies the conclusion. If we can do so, we will have shown that the argument is invalid. On the other hand, if the argument is valid we will find that it is not possible to draw a diagram that agrees with both premises yet denies the conclusion.

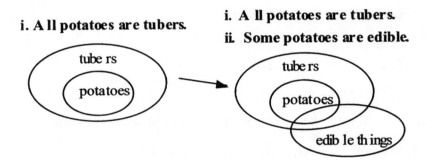

The diagram at right agrees with both premises and also agrees with the conclusion. This alone does not guarantee that the argument is valid. What is significant is that we recognize while drawing this diagram that, due to the way that the sets are nested, **every** diagram that agrees with both premises will **automatically** agree with the conclusion. (Because the set "potatoes" is contained within the set "tubers," and the set "edible things" must overlap the set "potatoes," the set "edible things" cannot avoid overlapping the set "tubers.") Since it is not possible to draw a diagram that agrees with both premises but denies the conclusion, this argument is **valid.**

B. All vipers are snakes and some reptiles aren't snakes. Therefore, some reptiles aren't vipers.

In this case, the argument is equivalent to:

All vipers are snakes. Some reptiles aren't snakes. Therefore, some reptiles aren't vipers.

We will try to draw a diagram that agrees with both premises but denies the conclusion.

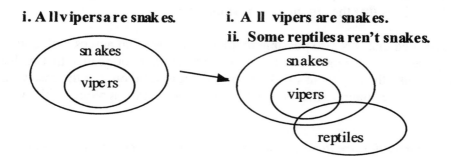

The diagram at right agrees with both premises and also agrees with the conclusion. This alone does not guarantee that the argument is valid. What is significant is that we recognize while drawing this diagram that, due to the way that the sets are nested, **every** diagram that agrees with both premises will **automatically** agree with the conclusion. (Because the set "vipers" is contained within the set "snakes," and at least some part of the set "reptiles" must be outside of the set "snakes," that same part of the set "reptiles" will also be outside of the set "vipers.") Since it is not possible to draw a diagram that agrees with both premises but denies the conclusion, this argument is **valid.**

C. Some motorcycles are noisy vehicles and some motorcycles are fast vehicles. Therefore, some noisy vehicles are fast vehicles.

This argument is equivalent to:

Some motorcycles are noisy vehicles. Some motorcycles are fast vehicles. Therefore, some noisy vehicles are fast vehicles.

We will try to draw a diagram that agrees with both premises but denies the conclusion.

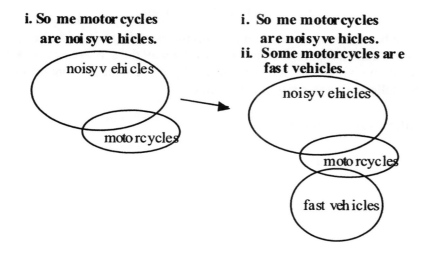

We have drawn a diagram that agrees with both premises but does not agree with the conclusion. This tells us that the argument is **invalid.**

You may have noticed that it is also possible to draw a diagram that agrees with both premises and agrees with the conclusion, too. It is essential that you realize that such a diagram does not make the argument valid. An argument is valid only in the case where it is impossible to draw a diagram that denies the conclusion while agreeing with every premise.

D. All coastal communities are flood-prone and Carrabelle is a coastal community. Therefore, Carrabelle is flood-prone.

This argument is equivalent to:

All coastal communities are flood-prone. Carrabelle is a coastal community. Therefore, Carrabelle is flood-prone.

We will try to draw a diagram that agrees with both premises but denies the conclusion.

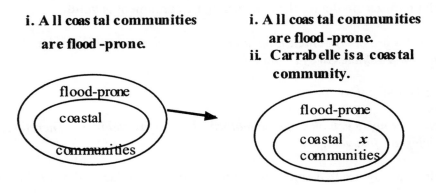

i. All coastal communities are flood-prone.

i. All coastal communities are flood-prone.
ii. Carrabelle is a coastal community.

Let x = Carrabelle

Again, the diagram we have drawn agrees with both premises and agrees with the conclusion. However, we must realize that, because of the way the sets are nested, **every** diagram that agrees with both premises will **automatically** agree with the conclusion. It is this realization that tells us that this argument is **valid.**

The correct choice (the invalid argument) is C.

EXAMPLE B

All of the arguments A - D have true conclusions, but one of the arguments is not valid. Select the argument that is **not** valid.

A. If an animal is an insect, then it has six legs. Beetles are insects. Therefore, beetles have six legs.
B. If one is a pitcher, then one is a baseball player. Shaq is not a pitcher. Therefore, Shaq is not a baseball player.
C. If a book is a novel, then it is fiction. My math text is non-fiction. Therefore, my math text is not a novel.
D. If today is the winter solstice, then today is the shortest day of the year. If today is the shortest day of the year, then this month is December. Therefore, if today is the winter solstice, then this month is December.

EXAMPLE B SOLUTION

We will test the validity of each argument.

A. If an animal is an insect, then it has six legs. Beetles are insects. Therefore, beetles have six legs.

Let *p* be "An animal is an insect."   Let *q* be "An animal has six legs."

The argument has this form:

$p \rightarrow q$

$\underline{p \qquad}$

$\therefore q$

(Note: in the second premise and conclusion we are letting the specific subject "beetles" take the place of the general subject "an animal" from the first premise.) We should recognize that this is a **valid** argument, because it is an example of **Direct Reasoning.**

B. If one is a pitcher, then one is a baseball player. Shaq is not a pitcher. Therefore, Shaq is not a baseball player.

Let *p* be "One is a pitcher."   Let *q* be "One is a baseball player."

The argument has this form:

$p \rightarrow q$

$\underline{\sim p \qquad}$

$\therefore \sim q$

(Note: in the second premise and conclusion we are letting the specific subject "Shaq" take the place of the general subject "one" from the first premise.)

We should recognize that this argument is **invalid,** because it is an example of **Fallacy of the Inverse.**

C. If a book is a novel, then it is fiction. My math text is non-fiction. Therefore, my math text is not a novel.

Let *p* be "A book is a novel."   Let *q* be "A book is fiction."

The argument has this form:

$p \rightarrow q$

$\underline{\sim q \quad\quad}$

$\therefore \sim p$

(Note: in the second premise and conclusion we are letting the specific subject "my math text" take the place of the general subject "a book" from the first premise.)

We should recognize that this is a **valid** argument, because it is an example of **Contrapositive Reasoning.**

D. If today is the winter solstice, then today is the shortest day of the year. If today is the shortest day of the year, then this month is December. Therefore, if today is the winter solstice, then this month is December.

Let *p* be "Today is the winter solstice." Let *q* be "Today is the shortest day of the year." Let *r* be "This month is December."

The argument has this form:

$p \rightarrow q$

$\underline{q \rightarrow r}$

$\therefore p \rightarrow r$

We should recognize that this is a **valid** argument, because it is an example of **Transitive Reasoning.**

The correct choice (the invalid argument) is B.

For more practice with this skill go to pages 183 and 188 of this manual.

CLAST SKILL III.E.1
The student recognizes valid reasoning patterns as illustrated by valid arguments in everyday language.

Refer to Section 3.6 of *Thinking Mathematically*.

You will be given two premises and asked to select the conclusion that will result in a valid argument. Unlike other CLAST skills involving arguments, this skill will only involve valid arguments.

The arguments will be restricted to these forms:

Direct Reasoning	Contrapositive Reasoning	Disjunctive Syllogism	Disjunctive Syllogism	Transitive Reasoning
$p \to q$	$p \to q$	$p \vee q$	$p \vee q$	$p \to q$
p	$\sim q$	$\sim p$	$\sim q$	$q \to r$
$\therefore q$	$\therefore \sim p$	$\therefore q$	$\therefore p$	$\therefore p \to r$

EXAMPLE A

Select the conclusion that will make the argument valid.

If you are too cheap to spend 35 cents, then you call collect.
You didn't call collect.

A. You are too cheap to spend 35 cents.
B. If you call collect, then you are too cheap to spend 35 cents.
C. You aren't too cheap to spend 35 cents.
D. You steal packets of mustard from fast food restaurants, too.

EXAMPLE A SOLUTION

Let p be "You are too cheap to spend 35 cents."   Let q be "You call collect."

The premises have this symbolic arrangement:

$p \to q$

$\sim q$

We should recognize this as the premise arrangement for **Contrapositive Reasoning**, which is a form of valid reasoning. This means that we will be able to form a valid conclusion according to this pattern:

SKILL III.E.1

$$p \rightarrow q$$
$$\underline{\sim q}$$
$$\therefore \sim p$$

In words, the valid conclusion is "You aren't too cheap to spend 35 cents."

The correct choice is C.

EXAMPLE B

Select the conclusion that will make the argument valid.

Class is cancelled or some of my classmates are here.
None of my classmates is here.

A. Class isn't cancelled.
B. If class is cancelled then some of my classmates aren't here.
C. This happens every Monday after a test.
D. Class is cancelled.

EXAMPLE B SOLUTION

Let p be "Class is cancelled."   Let q be "Some of my classmates are here."

The premises have this symbolic arrangement:

$$p \lor q$$
$$\sim q$$

(Recall that the negation of "Some of my classmates are here" is "None of my classmates is here.")

We should recognize this as the premise arrangement for **Disjunctive Syllogism**, which is a form a valid reasoning. This means that we will be able to form a valid conclusion by using this pattern:

$$p \lor q$$
$$\underline{\sim q}$$
$$\therefore p$$

In words, the valid conclusion is "Class is cancelled."

The correct choice is D.

For more practice with this skill go to page 184 of this manual.

CLAST SKILL III.E.2
The student selects applicable rules for transforming statements without affecting their meaning.

Refer to Sections 3.1 and 3.5 of *Thinking Mathematically.*

You will be given a pair of simple or compound statements expressed in words. The two statements will be equivalent, and you will be asked to select the rule that establishes the equivalence of the statements.

The correct rule of equivalency may be any of these:

$p \rightarrow q \equiv {\sim}p \vee q$

$p \rightarrow q \equiv {\sim}q \rightarrow {\sim}p$

${\sim}({\sim}p) \equiv p$

${\sim}(p \wedge q) \equiv {\sim}p \vee {\sim}q$

${\sim}(p \vee q) \equiv {\sim}p \wedge {\sim}q$

${\sim}(p \rightarrow q) \equiv p \wedge {\sim}q$

${\sim}$(All are p) $\equiv$ Some are ${\sim}p$

${\sim}$(Some are p) $\equiv$ All are ${\sim}p$

${\sim}$(Some are p) $\equiv$ None are p

The rules will be stated in words, rather than symbols.

You could probably solve one of these problems even if you didn't know any of those rules of equivalency, since choosing the correct answer depends primarily upon your ability to recognize which one of the choices has the same pattern as the two given statements.

Some of the incorrect choices may list rules of equivalence that do not apply to the given pair of statements, while other incorrect choices may list "rules" that are logically incorrect.

SKILL III.E.2

EXAMPLE

Select the rule of logical equivalence which directly (in one step) transforms statement "i" into statement "ii."

i. If x is an odd number, then $3x$ is an odd number.

ii. If $3x$ is not an odd number, then x is not an odd number.

A. "Not (p or q) is equivalent to "not p and not q."

B. "If p, then q" is equivalent to "If not q, then not p."

C. "If p, then q" is equivalent to "If not p, then not q."

D. Correct equivalence is not given.

SOLUTION

Let p be the statement "x is an odd number."

Let q be the statement "$3x$ is an odd number."

Statement "i" corresponds to "If p, then q."

Statement "ii" corresponds to "If not q, then not p."

Choice B correctly establishes the equivalence between these two statements.

Notice that choice A is a correct rule of logical equivalence, but it has nothing to do with the particular statements in this problem.

Notice also that choice C is a "rule" that is logically incorrect.

For more practice with this skill go to page 179 of this manual.

CLAST SKILL IV.E.1 *The student draws logical conclusions when the facts warrant them.*

Refer to Sections 3.6 and 3.7 of *Thinking Mathematically.*

You will be given from two to four premises and asked to select the conclusion "that is warranted," if such a conclusion is listed. This means that you must select the conclusion that would produce a valid argument when combined with the given premises. In some cases, it may not be possible to form a non-trivial conclusion from the given premises. In other cases, even though it is possible to form a valid conclusion, that conclusion might not be among the choices that are given.

It will be useful to recall these common patterns of **valid** reasoning.

Direct Reasoning	Contrapositive Reasoning	Disjunctive Syllogism	Disjunctive Syllogism	Transitive Reasoning
$p \rightarrow q$	$p \rightarrow q$	$p \vee q$	$p \vee q$	$p \rightarrow q$
p	$\sim q$	$\sim p$	$\sim q$	$q \rightarrow r$
$\therefore q$	$\therefore \sim p$	$\therefore q$	$\therefore p$	$\therefore p \rightarrow r$

It will also be useful to recognize these common patterns of **invalid** reasoning:

Fallacy of the Converse	Fallacy of the Inverse	Misuse of Disjunction	Misuse of Transitive Reasoning	Misuse of Transitive Reasoning
$p \rightarrow q$	$p \rightarrow q$	$p \vee q$	$p \rightarrow q$	$p \rightarrow q$
q	$\sim p$	p	$p \rightarrow r$	$r \rightarrow q$
$\therefore p$	$\therefore \sim q$	$\therefore \sim q$	$\therefore q \rightarrow r$	$\therefore p \rightarrow r$

Some of the arguments presented in this skill may be analyzed with Euler diagrams, as well.

161

SKILL IV.E.1

EXAMPLE A

Study the information given below. If a logical conclusion is given, select that conclusion. If none of the conclusions given is warranted, select the option expressing this condition.

If one is a figure skater, then one is graceful. If one is graceful, then one does not stomp. Sally stomps.

A. Sally isn't a figure skater.
B. If one isn't a figure skater, then one stomps.
C. Sally is a figure skater.
D. None of the above is warranted.

EXAMPLE A SOLUTION

Let p be the statement "One is a figure skater."
Let q be the statement "One is graceful." Let r be the statement "One doesn't stomp."

The premises have this symbolic form:

$p \rightarrow q$

$q \rightarrow r$

$\sim r$

(Note: in the third premise we are using the specific subject "Sally" in place of the general subject "one" from the other premises.)

We must first recognize that the first two premises conform to the premise arrangement for **Transitive Reasoning**. Those two premises combined are equivalent to $p \rightarrow r$. Thus, we can reduce the premise scheme by replacing the first two premises with $p \rightarrow r$. The reduced premise scheme has this form:

$p \rightarrow r$

$\sim r$

We need to recognize that this fits the premise arrangement for **Contrapositive Reasoning**. We will be able to form a valid conclusion by referring to this pattern:

$p \rightarrow r$

$\underline{\sim r}$

$\therefore \sim p$

In words, the valid conclusion is "Sally isn't a figure skater."

The correct choice is A.

162

EXAMPLE B

Study the information given below. If a logical conclusion is given, select that conclusion. If none of the conclusions given is warranted, select the option expressing this condition.

If you turn down the thermostat, you will use less electricity.
If you wash your clothes in cold water, you will use less electricity.

A. If you use less electricity, then you turn down your thermostat.
B. If you don't wash you clothes in cold water, then you won't use less electricity.
C. If you turn down the thermostat, then you wash your clothes in cold water.
D. None of the above is warranted.

EXAMPLE B SOLUTION

Let p be the statement "You turn down the thermostat."
Let q be the statement "You will use less electricity."
Let r be the statement "You wash your clothes in cold water."

The premises have this symbolic arrangement:

$p \rightarrow q$

$r \rightarrow q$

This does not agree with the premise arrangement for Transitive Reasoning, and that fact suggests that we may not be able to arrive at a non-trivial valid conclusion. The worst-case scenario for a problem like this, in which a common pattern of valid reasoning does not occur, is that we may need to make three truth tables in order to test the validity of three different arguments (one argument for each of the three different conclusions given in choices A, B and C.)

If we choose C for the conclusion, we have the following pattern:

$p \rightarrow q$

$\underline{r \rightarrow q}$

$\therefore p \rightarrow r$

As mentioned above, this is a common pattern for invalid reasoning, which is sometimes referred to as a Misuse of Transitive Reasoning. For this reason, choice C is not correct.

Choice B is not a valid conclusion, either. We can explain why without having to make a truth table. The statement in choice B is the inverse of the second premise. We cannot

SKILL IV.E.1

use a conditional statement to logically deduce its inverse; to do so results in a fallacy that is similar to Fallacy of the Inverse.

For a similar reason, choice A is not a valid conclusion. The statement in choice A is the converse of the first premise. If we try to use a conditional statement to deduce its converse, we are committing a logical error that is similar to Fallacy of the Converse.

The claims that both choices A and B are not valid conclusions can be verified with truth tables.

The correct choice is D.

EXAMPLE C

Study the information given below. If a logical conclusion is given, select that conclusion. If none of the conclusions given is warranted, select the option expressing this condition.

All rodents love to gnaw. All nutria are rodents. Sylvester does not love to gnaw.

A. Sylvester is a rodent.
B. Some nutria don't love to gnaw.
C. Sylvester isn't a nutria.
D. None of the above is warranted.

EXAMPLE C SOLUTION

We will use Euler diagrams to analyze the argument.

If a valid conclusion is warranted, we should find that every time we draw a diagram that agrees with every premise, the diagram will automatically agree with that conclusion.

On the other hand, if a conclusion is not warranted, we should find it possible to draw a diagram that agrees with every premise but denies the unwarranted conclusion.

We start with a diagram that agrees with the first premise. Such a diagram must show that the set of "rodents" is contained within the set of things that "love to gnaw."

i. All rodents love to gnaw.

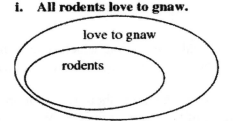

Now we expand the diagram so that it agrees with the second premise, too. The expanded diagram must show the set of "nutria" contained within the set of "rodents."

    i. All rodents love to gnaw.
   ii. All nutria are rodents.

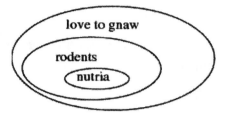

Now we expand the diagram so that it agrees with the third premise. This is a statement about an individual, so we will use the symbol x to represent that individual. According to the third premise, Sylvester must be placed outside of the set of things that "love to gnaw."

    i. All rodents love to gnaw.
   ii. All nutria are rodents.
  iii. Sylvester doesn't love to gnaw.

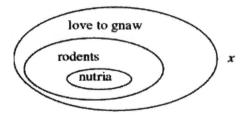

Let x = "Sylvester"

Notice that because of the way the sets are nested, the diagram will automatically reflect a number of valid conclusions, including:

All nutria love to gnaw.
Sylvester isn't a rodent.
Sylvester isn't a nutria.

Choice C is a valid conclusion.

For more practice with this skill go to pages 186 and 189 of this manual.

PART II
CLAST EXERCISES

CLAST Exercises for Section 1.1 of *Thinking Mathematically.*

CLAST SKILL III.A.1 *The student infers relations between numbers in general by examining particular number pairs.*

See page 21 for information about this CLAST skill.

1. Look for a common linear relationship between the numbers in each pair. Then identify the missing term.
(5, 10) (–10, –5) (0, 5) (1.2, 6.2) (1/5, 26/5) (15, ___)
A. 3   B. 10   C. 20   D. 30

2. Look for a common linear relationship between the numbers in each pair. Then identify the missing term.
(12, 9) (6, 3) (5.2, 2.2) (13/4, 1/4) (3, 0) (___, 3)
A. 3   B. –3   C. 0   D. 6

3. Look for a common linear relationship between the numbers in each pair. Then identify the missing term.
(12, 3) (16, 4) (1.6, .4) (1, 1/4) (0, 0) (–4, ___)
A. –1   B. –13   C. 5   D. 1

4. Look for a common linear relationship between the numbers in each pair. Then identify the missing term.
(–2, –4) (2, 4) (1/2, 1) (2/3, 4/3) (3.1, 6.2) (–2.5, ___)
A. –4.5   B. –5   C. 5   D. –.5

5. Look for a common quadratic relationship between the numbers in each pair. Then identify the missing term.
(–2, 4) (2, 4) (1/2, 1/4) (2.5, 6.25) (3, 9) (–3, ___)
A. –9   B. –18   C. 3   D. 9

6. Look for a common quadratic relationship between the numbers in each pair. Then identify the missing term.
(0, 0) (1, 1) (1, –1) (25, 5) (9, 3) (___, 4)
A. 0   B. 2   C. 16   D. –2

CLAST Exercises for Section 1.2 of *Thinking Mathematically.*

CLAST SKILL II.A.5 *The student will identify a reasonable estimate of a sum, average or product of numbers.*

See page 18 of this book to learn about this CLAST skill.

1. An investor owns 38.2 shares of a stock that is valued at $42 $\frac{1}{8}$ per share. Which of the following could be a reasonable estimate of the total value of the stock?

A. $1,200 B. $2,500 C. $1,600 D. $1,000

2. A shopper purchases three cans of soup at 79¢ each, two pounds of ground beef at $1.59 per pound, a loaf of bread at $1.39, and a head of cauliflower at $2.29. Which of the following is a reasonable estimate of the total cost of these items?

A. $9 B. $12 C. $5 D. $15

3. Gerry is going to buy a computer, printer, scanner, software, and mass storage drive. The prices for these items are $1,444.99, $129.99, $79.99, $89.99 and $179.99 respectively, plus an additional $119 to cover shipping and an extended warranty for the entire order. Which of the following is a reasonable estimate of the total cost of this purchase?

A. $2,100 B. $2,800 C. $3,500 D. $3,100

4. Andrea bought 12 CD recordings. The least expensive recording was $6.99 and the most expensive was $15.99. Which of the following is a reasonable estimate of the total cost of this purchase?

A. $130 B. $84 C. $80 D. $190

5. There are 33 students in a kindergarten class. For their class picnic, 21 students order a pizza meal for $1.89 and the rest order a hamburger meal for $1.19 apiece. Which of the following is a reasonable estimate of the total cost of this purchase?

A. $50 B. $66 C. $85 D. $33

6. A restaurant manager has ordered 29 cases of frozen, cut-up chicken. The weights of each case vary from 49 pounds to 55 pounds. Which of the following is a reasonable estimate of the total weight?

A. 15,000 lb. B. 1,500 lb. C. 1,800 lb. D. 26,000 lb.

7. In a math class of thirty students, the student with the lowest math SAT score had a 440, and the student with the highest math SAT score had a 610. Which of the following is a reasonable estimate of the average SAT score for this group of students?

A. 400   B. 15,000   C. 1,500   D. 510

8. There are 40 dogs at the animal shelter. The smallest dog weighs 8 pounds and the largest dog weighs 140 pounds. Which of the following is a reasonable estimate of the average weight?

A. 140 pounds   B. 2000 pounds   C. 24 pounds   D. 6 pounds

CLAST SKILL I.D.1 *The student will identify information contained in bar, line and circle graphs.*

See page 124 of this book to learn about this CLAST skill.

9. The line graph below shows the total number of base hits collected by a youth league baseball player after each of the first seven games of the season. During which game did the player not get any hits?

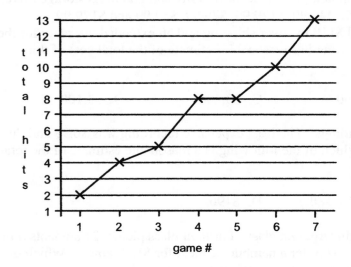

A. Game #1

B. Game #2

C. Game #5

D. Game #7

168

10. The bar graph below shows the distribution of scores on an exam.
For which score is it true that 28 students had scores greater than that score?

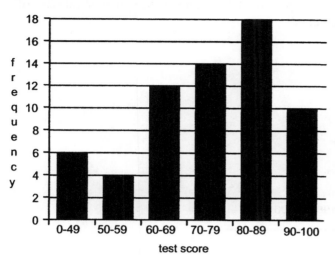

A. 89   B. 79   C. 14   D. 18

11. The bar graph below shows the total U.S. spending on advertising for various categories. What was the total combined advertising expenditure for the Drugs & Remedies and Toiletries/Cosmetics categories?

A. $4 billion   B. $3.8 billion   C. $7.8 billion   D. $13 billion

1997 Advertising Expenditure (in $ billions)

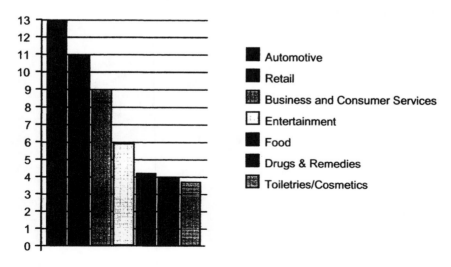

(Source: Competitive Media Reporting and Publishers Information Bureau)

SECTION 1.2

12. The graph below conveys information about the air quality in Los Angeles, CA. For each of the years 1987 – 1996 the graph shows the number of days on which the air quality was below EPA standards. Approximately what was the greatest number of substandard days in any of these years?

A. 200

B. 300

C. 240

D. 88

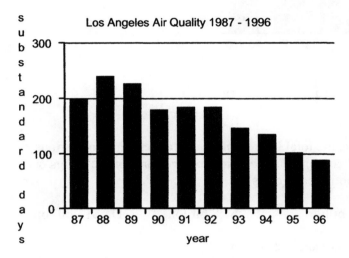

(Source: Environmental Protection Agency)

CLAST Exercises for Sections 2.2 and 2.3 of *Thinking Mathematically*

CLAST Skill I.E.1 *The student will deduce facts about set-inclusion or non-set-inclusion from a diagram.*

See page 140 of this book to learn about this CLAST skill.

1. Sets S, T, U and V are related as shown in the diagram. Which of the following statements is true, assuming that none of the regions of the diagram is empty?

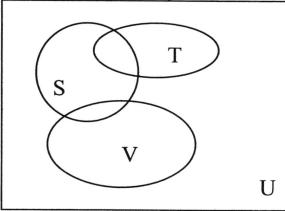

A. Any element that is a member of S is also a member of T.
B. There is no element that is a member of both S and V.
C. There is at least one element that is a member of both T and S.
D. None of the above is true.

2. Sets U, V, W and X are related as shown in the diagram. Which of the following statements is true, assuming that none of the regions of the diagram is empty?

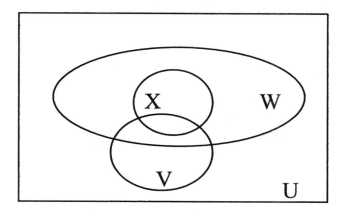

A. There is no element common to sets V, X and W.
B. Any element that is a member of both sets X and V is also a member of set W.
C. Any element that is a member of U is also a member of set X or set V or set W.
D. None of the above is true.

3. Sets U, A, B and C are related as shown in the diagram. Which of the following statements is true, assuming that none of the regions of the diagram is empty?

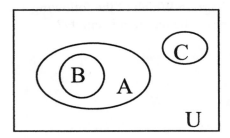

A. There is no element common to sets A, B and C.
B. Any element that is a member of set A is also a member of set B.
C. Any element that is not a member of set C is a member of set A.
D. None of the above is true.

4. Sets S, T, U and V are related as shown in the diagram. Which of the following statements is true, assuming that none of the regions of the diagram is empty?

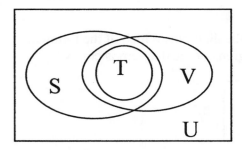

A. Any element that is a member of sets S and V is also a member of set T.
B. There is no element common to sets S, T and V.
C. Any element that is not a member of set V is not a member of set T.
D. None of the above is true.

5. Sets A, B, C and U are related as shown in the diagram. Which of the following statements is true, assuming that none of the regions of the diagram is empty?

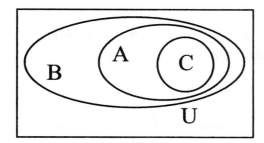

A. Any element that is a member of set A is also a member of set C.
B. There is no element common to sets B and C.
C. There is at least one element that is a member of sets A and C but not of set B.
D. None of the above is true.

6. Sets S, V, T, and U are related as shown in the diagram.
Which of the following statements is true, assuming that none of the regions of the diagram is empty?

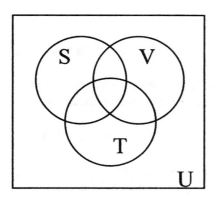

A. Any element which is a member of V is also a member of S.
B. No element is a member of both sets V and T.
C. Any element that is a member of sets S and T is also a member of set V.
D. None of these is true.

CLAST Exercises for Section 3.1 of *Thinking Mathematically.*

CLAST Skill II.E.1 *The student will identify statements equivalent to the negations of simple and compound statements.*

See page 142 of this book to learn about this CLAST skill.

1. Select the statement that is the negation of the statement "All swans are stately."
A. Some swans are stately.
B. No swans are stately.
C. All states are swanly.
D. Some swans aren't stately.

2. Select the statement that is the negation of the statement "Some poodles chew shoes."
A. All poodles chew shoes.
B. No poodles chew shoes.
C. If it chews shoes, then it isn't a poodle.
D. Some poodles don't chew shoes.

3. Select the statement that is the negation of the statement "Some senators are running for re-election."
A. Some senators aren't running for re-election.
B. At least one senator is running for re-election.
C. No senators are running for re-election.
D. All senators are running for re-election.

4. Select the statement that is the negation of the statement "All mountain towns are quiet."
A. No mountain towns are quiet.
B. Some mountain towns aren't quiet.
C. Some mountain towns are quiet.
D. All quiet towns are mountain towns.

5. Select the statement that is the negation of the statement "Some poems don't make sense."
A. All poems make sense.
B. No poems make sense.
C. Some poems make sense.
D. Some poets don't make cents.

6. Select the statement that is the negation of the statement "All of my shoes are uncomfortable."
A. None of my shoes is uncomfortable.
B. None of my shoes is comfortable.
C. Some of my shoes are uncomfortable.
D. Some of my shoes are comfortable.

7. Select the statement that is the negation of the statement "Some reindeer can fly."
A. No reindeer can fly.
B. All reindeer can fly.
C. Some reindeer can't fly.
D. Some flies eat canned reindeer.

8. Select the statement that is the negation of the statement "All of my in-laws are in jail."
A. None of my in-laws is in jail.
B. Some of my in-laws are in jail.
C. Some of my in-laws aren't in jail.
D. If somebody is in jail, then he or she is one of my in-laws.

9. Select the statement that is the negation of the statement "All used cars are unreliable."
A. No used cars are reliable.   B. All used cars are reliable.
C. Some used cars are unreliable.   D. Some used cars are reliable.

10. Select the statement that is the negation of the statement "Some politicians are dishonest."
A. All politicians are honest.   B. No politicians are honest.
C. Some politicians are honest.   D. Some politicians are dishonest.

CLAST SKILL III.E.2 *The student selects applicable rules for transforming statements without affecting their meaning.*

See page 159 of this book to learn about this CLAST skill.

11. Select the rule of logical equivalence which directly (in one step) transforms statement "i" into statement "ii."

i. Not all of my puppies are trained.
ii. Some of my puppies are not trained.

A. "If p, then q" is equivalent to "(not p) or q."
B. "Not all are p" is equivalent to "some are not p."
C. "If p, then q" is equivalent to "If not p, then not q."
D. "Not (not p) is equivalent to "p."

12. Select the rule of logical equivalence which directly (in one step) transforms statement "i" into statement "ii."

i. It is not the case that some thunderstorms are pleasant.
ii. All thunderstorms are unpleasant.

A. "Some are p" is equivalent to "some aren't p."
B. "Not all are p" is equivalent to "all are (not p)."
C. "Not (some are p)" is equivalent to "all are (not p)."
D. Correct equivalence rule is not given.

SECTION 3.5

CLAST Exercises for Section 3.5 of *Thinking Mathematically.*

CLAST Skill II.E.1 *The student will identify statements equivalent to the negations of simple and compound statements.*

See page 142 of this book to learn about this CLAST skill.

1. Select the statement that is the negation of the statement "I am taking a math class and an English class."

A. I am taking a math class or an English class.
B. I am not taking a math class and I am not taking an English class.
C. I am not taking a math class or I am not taking an English class.
D. I am taking a math class or I am not taking an English class.

2. Select the statement that is the negation of the statement "The poppy field will make you sleep or the flying monkeys will carry you away."

A. The poppy field won't make you sleep or the flying monkeys won't carry you away.
B. The poppy field won't make you sleep and the flying monkeys won't carry you away.
C. The poppy field will make you sleep and the flying monkeys won't carry you away.
D. The poppy field will make you sleep or the flying monkeys won't carry you away.

3. Select the statement that is the negation of "If you like *film noir*, then you like The *Big Sleep.*"

A. You don't like *film noir* and you like *The Big Sleep.*
B. You like *film noir* and you don't like *The Big Sleep.*
C. If you don't like *film noir,* then you don't like *The Big Sleep.*
D. If you don't like *The Big Sleep,* then you don't like *film noir.*

4. Select the statement that is the negation of this statement (from the instructions for IRS Form 1040): "If you did not pay enough, we will send you a bill."

A. If you did not pay enough, we won't send you a bill.
B. You did not pay enough or we will send you a bill.
C. You did not pay enough and we won't send you a bill.
D. If you paid enough, then we won't send you a bill.

5. Select the statement that is the negation of the statement "You have a credit card or our date is cancelled."

A. If you have a credit card then our date isn't cancelled.
B. You don't have a credit card or our date isn't cancelled.
C. You don't have a credit card and our date isn't cancelled.
D. You don't have a credit card and our date is cancelled.

CLAST SKILL II.E.2 *The student will determine equivalence or non-equivalence of statements.*

See page 145 of this book to learn about this CLAST skill.

6. Select the statement that is logically equivalent to "If I will retire in comfort, then I will start planning today."

A. If I won't retire in comfort, then I won't start planning today.
B. I will retire in comfort and I won't start planning today.
C. If I will start planning today, then I will retire in comfort.
D. If I won't start planning today, then I won't retire in comfort.

7. Select the statement that is logically equivalent to "It is not true that if you pass this course then you will graduate."

A. You pass this course and you won't graduate.
B. You won't pass this course or you will graduate.
C. If you don't pass this course then you won't graduate.
D. If you don't pass this course then you will graduate.

8. Select the statement that is logically equivalent to "If you aren't a careful driver, then you'll wreck your car."

A. If you are a careful driver, then you won't wreck your car.
B. If you wreck your car, then you aren't a careful driver.
C. You are a careful driver or you wreck your car.
D. You are a careful driver and you don't wreck your car.

9. Select the statement that is **not** logically equivalent to this statement (from the instructions for IRS Form 1040): "If you have paid too much, we will send you a refund."

A. If we won't send you a refund, then you haven't paid too much.
B. If you haven't paid too much, then we won't send you a refund.
C. You haven't paid too much or we will send you a refund.
D. We will send you a refund, or you haven't paid too much.

SECTION 3.5

10. Select the statement that is logically equivalent to "It is not true that both roses are red and violets are blue."

A. Roses are red and violets aren't blue.
B. Roses are red or violets aren't blue.
C. Roses aren't red and violets aren't blue.
D. Roses aren't red or violets aren't blue.

11. Select the statement that is **not** logically equivalent to "If you aren't gullible, then you won't vote for me."

A. If you won't vote for me, then you aren't gullible.
B. If you will vote for me, then you are gullible.
C. You won't vote for me, or you are gullible.
D. You are gullible, or you won't vote for me.

12. Select the statement that is logically equivalent to "It is not the case that if I don't get a better job then I don't get a better apartment."

A. If I get a better job, then I get a better apartment.
B. If I don't get a better job, then I get a better apartment.
C. I get a better job and I don't get a better apartment.
D. I don't get a better job and I get a better apartment.

13. Select the statement that is logically equivalent to "If you know the answer, then you don't guess."

A. You don't know the answer or you don't guess.
B. You know the answer and you guess.
C. If you don't know the answer, then you guess.
D. If you don't guess, then you know the answer.

CLAST SKILL III.E.2 *The student selects applicable rules for transforming statements without affecting their meaning.*

See page 159 of this book to learn about this CLAST skill.

14. Select the rule of logical equivalence which directly (in one step) transforms statement "i" into statement "ii."

i.   If you need a book, then you go to the library.
ii. If you don't go to the library, then you don't need a book.

A. "If p, then q" is equivalent to "(not p") or q."
B. "Not (p or q)" is equivalent to "not p or not q."
C. "If p, then q" is equivalent to "If not p, then not q."
D. Correct equivalence rule is not given.

15. Select the rule of logical equivalence which directly (in one step) transforms statement "i" into statement "ii."

i.   It is not true that today isn't Saturday.
ii. Today is Saturday.

A. "Not all are p" is equivalent to "some are not p."
B. "Not (not p)" is equivalent to "p."
C. "Not (p or q)" is equivalent to "not p and not q."
D. Correct equivalence rule is not given.

16. Select the rule of logical equivalence which directly (in one step) transforms statement "i" into statement "ii."

i.   It is not the case that my house has both termites and a leaky roof.
ii. My house doesn't have termites or my house doesn't have a leaky roof.

A. "Not (p or q)" is equivalent to "not p and not q."
B. "Not (p and q)" is equivalent to "not p and not q."
C. "Not (p and q)" is equivalent to "not p or not q."
D. "Not (p or q)" is equivalent to "not p or not q."

CLAST Exercises for Section 3.6 of Thinking *Mathematically*

CLAST Skill II.E.3 *The student will draw logical conclusions from data.*

See page 148 of this book for information about this CLAST skill.

1. Given that:

i. If you hit a home run, then you get an RBI;
ii. You didn't get an RBI.

determine which conclusion can be logically deduced.

A. You didn't hit a home run.
B. You hit a home run.
C. You scored a touchdown.
D. None of the above.

2. Given that:

i. This animal has a shell or it isn't a turtle.
ii. This animal is a turtle.

determine which conclusion can be logically deduced.

A. This animal is likes to swim.
B. This animal has a shell.
C. This animal doesn't have a shell.
D. None of the above.

3. Given that:

i. If this motorcycle is a Harley-Davidson, then I can't afford it.
ii. This motorcycle isn't a Harley-Davidson.

determine which conclusion can be logically deduced.

A. I can't afford it.
B. I need to buy a leather jacket.
C. I can afford it.
D. None of the above.

4. Study the requirements <u>and</u> each applicant's qualifications for a debt consolidation loan of $15,000. Then identify which of the applicants would qualify for the loan.

To qualify for a loan of $15,000 the applicant must have a gross annual income of at least $25,000 if single ($40,000 combined income if married), must have lived at his/her current address for at least 2 years, and must have an outstanding credit card balance of not less than $7,500 and not more than $20,000.

Mr. Brown is single, has a gross annual income of $45,000, has lived in his current residence for 3 years, and has an outstanding credit card balance of $7,000.

Ms. Smith is married and has a gross annual of income of $28,000. Her husband's gross annual income is $23,000. They have lived in their current residence for 1 year, and have an outstanding credit card balance of $14,000.

Mr. Allen is married and has a gross annual income of $19,000. His wife's gross annual income is $17,000. They have lived in their current residence for 4 years and have an outstanding credit card balance of $12,000.

A. Mr. Brown
B. Ms. Smith
C. Mr. Allen
D. No one is qualified.

5. Read the requirements <u>and</u> each applicant's qualifications for transfer admission to the School of Business at Major University. Then identify which of the applicants would qualify for admission.

To qualify for admission the applicant must

I. Have an Associate's Degree; and

II. Have passed, with grades of C- or better, a two-semester or three-quarter college sequence in a single foreign language, or two years of high school credit in a single foreign language; and

III. Have credit with grades of C- or better in the courses Calculus for Business and Statistics for Business, or have a cumulative grade point average of 3.50 (on a 4.0 scale) or better.

Anna has an AA, has college grades of A in both Spanish I and French I, grades of B and B+ respectively in Calculus for Business and Statistics for Business, and has a cumulative grade point average of 3.78.

SECTION 3.6

Charles has an AA, has college grades of C+ and B respectively in the two-semester sequence German I and German II, has a grade of A- in Calculus for Business, has not taken Statistics for Business, and has a cumulative grade point average of 3.10.

Mark has an AA, two years of high school French, grades of C and B- respectively in Calculus for Business and Statistics for Business, and a cumulative grade point average of 2.48.

A. Anna
B. Charles
C. Mark
D. No one is qualified.

6. In 1999, people who paid interest on a qualified student loan were directed to use the following instructions to determine eligibility to deduct the interest payments from their Federal income tax.

Read the requirements (adapted from the instructions for IRS form 1040), and then determine which taxpayer is eligible for the deduction.

I. Your filing status is any status except married filing separately; and
II. Your modified adjusted gross income (AGI) is less than: $50,000 if single, head of household, or qualifying widow(er); $75,000 if married filing jointly; and
III. You are not claimed as a dependent on someone else's tax return.

Sally's filing status is head of household, her AGI is $18,000, and she is not claimed as a dependent on anybody else's tax return;

Jessica's filing status is single, her AGI is $53,000, and she is not claimed as a dependent on anybody else's tax return.

Sarah's filing status is single, her AGI is $12,000, and she is claimed as a dependent on her father's tax return.

A. Sally
B. Jessica
C. Sarah
D. None of them is eligible.

7. Given that:

i. If one is a clown, then one wears face paint;
ii. Bernadette isn't a clown;

determine which conclusion can be logically deduced.

A. Bernadette wears face paint.
B. Bernadette doesn't wear face paint.
C. Bernadette wears too much eye shadow.
D. None of the above.

CLAST SKILL II.E.4 *The student will recognize that an argument may not be valid even though its conclusion is true.*

See page 151 of this book for information about this CLAST skill.

8. All of the arguments A - D have true conclusions, but one of the arguments is not valid. Select the argument that is **not** valid.

A. If one is a horse, then one has hooves. An appaloosa is a horse. Therefore, an appaloosa has hooves.
B. If one is a frog, then one swims. Lead weights don't swim. Therefore, lead weights aren't frogs.
C. If one is a successful athlete, then one must have excellent stamina. If one has excellent stamina, then one is in good health. Therefore, if one is a successful athlete, then one is in good health.
D. If one is a snake, then one slithers. Pythons slither. Therefore, pythons are snakes.

9. All of the arguments A - D have true conclusions, but one of the arguments is not valid. Select the argument that is **not** valid.

A. If today is federal holiday, then the Post Office is closed. If the Post Office is closed, then I don't receive mail. Therefore, if today is a federal holiday, then I don't receive mail.
B. If an animal is a squirrel, then it has a bushy tail. An alligator is not a squirrel. Therefore, an alligator does not have a bushy tail.
C. If a film is rated G, then it is suitable for young children. *Mary Poppins* is rated G. Therefore, *Mary Poppins* is suitable for young children.
D. If you get too many moving violations, then your license is suspended. If your license is suspended, then you can't drive legally. Therefore, if you get too many moving violations, then you can't drive legally.

10. All of the arguments A - D have true conclusions, but one of the arguments is not valid. Select the argument that is **not** valid.

A. If an animal is a reptile, then it is cold-blooded. A penguin is not a reptile. Therefore, a penguin is not cold-blooded.
B. If you strike out, then you don't have a base hit. If you don't have a base hit, then your batting average does not increase. Therefore, if you strike out, then your batting average does not increase.
C. If an investment is a blue chip stock, then it is a sound investment. A "penny stock" is not a sound investment. Therefore, a "penny stock" is not a blue chip stock.
D. If you can't make a twenty- percent down payment, then you need mortgage insurance. If you need mortgage insurance, then your monthly payment increases. Therefore, if you can't make a twenty- percent down payment, then you monthly payment increases.

CLAST SKILL III.E.1 *The student recognizes valid reasoning patterns as illustrated by valid arguments in everyday language.*

See page 157 of this book for information about this CLAST skill.

11. Select the conclusion that will make the following argument valid.

You were once a worm or you aren't a butterfly. You weren't once a worm;

A. You aren't a butterfly.   B. You swallowed a worm.
C. You are a butterfly.   D. None of the above.

12. Select the conclusion that will make the following argument valid.

If I forget my password, then I can't check my e-mail. I can check my e-mail.

A. If I can't check my e-mail, then I forgot my password.
B. I forgot my password.
C. I didn't forget my password.
D. If I can't check my e-mail, then I didn't forget my password.

13. Select the conclusion that will make the following argument valid.

If an item is subject to sales tax, then it is not considered to be a food item. That snack is considered to be a food item.

A. That snack will spoil your appetite.
B. That snack is subject to sales tax.
C. That snack is bad for your teeth.
D. That snack is not subject to sales tax.

14. Select the conclusion that will make the following argument valid.

If a writer was a member of "The Lost Generation," then he or she was influenced by the First World War. If a writer was influenced by the First World War, then he or she did not die in the Nineteenth Century.

A. If a writer was a member of "The Lost Generation," then he or she did not die in the Nineteenth Century.
B. If a writer was not influenced by the First World War, then he or she died in the Nineteenth Century.
C. If a writer did not die in the Nineteenth Century, then he or she was a member of "The Lost Generation."
D. If a writer was influenced by the First World War, then he or she was a member of "The Lost Generation."

15. Select the conclusion that will make the following argument valid.

One is a congressperson or one isn't a United States Senator. Bob Graham is a United States Senator.

A. Bob Graham is a congressperson.
B. Bob Graham isn't a congressperson.
C. Bob Graham is a Democrat.
D. Bob Graham is a Republican.

16. Select the conclusion that will make the following argument valid.

If stock prices rise, then bond values fall. Stock prices are rising.

A. Bond values are rising.
B. James Bond should buy stocks.
C. Bond values are falling.
D. If bond values fall, then stock prices rise.

17. Select the conclusion that will make the following argument valid.

If some of my classes are difficult, then I don't have time to relax.
I have time to relax.

A. All of my classes are difficult.
B. Some of my classes are difficult.
C. None of my classes is difficult.
D. Some of my classes aren't difficult.

SECTION 3.6

CLAST SKILL IV.E.1 *The student draws logical conclusions when the facts warrant them.*

See page 161 of this book for information about this CLAST skill.

18. Study the information given below. If a logical conclusion is given, select that conclusion. If none of the conclusions given is warranted, select the option expressing this condition.

If one is a mule, then one is stubborn.   Francis is a mule.
If one is stubborn, then one is not easily persuaded.

A. If one is stubborn, then one is Francis.
C. Francis is not easily persuaded.

B. Francis is easily persuaded.
D. None of the above is warranted.

19. Study the information given below. If a logical conclusion is given, select that conclusion. If none of the conclusions given is warranted, select the option expressing this condition.

If you eat meat, then you aren't a vegetarian.
If you don't eat meat, then you don't eat pork.

A. If you aren't a vegetarian, then you eat pork.
B. If you are a vegetarian, then you don't eat pork.
C. If you eat vegetarians, then you are a cannibal.
D. None of the above is warranted.

20. Study the information given below. If a logical conclusion is given, select that conclusion. If none of the conclusions given is warranted, select the option expressing this condition.

If Joan's car has high insurance costs, she will get a second job. If Joan's car is a sports car, then it is expensive. Joan's car is a sports car. If Joan's car is expensive, then it has high insurance costs.

A. If Joan has a second job, then her car is a sports car.
B. If Joan doesn't have a second job, then her car isn't a sports car.
C. Joan will get a second job.
D. None of the above is warranted.

21. Study the information given below. If a logical conclusion is given, select that conclusion. If none of the conclusions given is warranted, select the option expressing this condition.
If I plant tomatoes, I will need fertilizer. If I plant peppers, I will need fertilizer.

A. If I plant tomatoes, then I will plant peppers.
B. If I plant peppers, then I will plant tomatoes.
C. If I don't plant tomatoes, then I won't need fertilizer.
D. None of the above is warranted.

22. Study the information given below. If a logical conclusion is given, select that conclusion. If none of the conclusions given is warranted, select the option expressing this condition.

If I paint my house, I will need a paintbrush. If I go to Hardware Henry's, then I will have to gas up the Pinto. If my house is hot pink, then I will paint it. If I need a paintbrush, then I will go to Hardware Henry's.

A. If I go to Hardware Henry's, then my house is hot pink.
B. If I don't need a paintbrush, then I won't have to gas up the Pinto.
C. If I don't have to gas up the Pinto, then my house isn't hot pink.
D. None of the above is warranted.

23. Study the information given below. If a logical conclusion is given, select that conclusion. If none of the conclusions given is warranted, select the option expressing this condition.
If you buy a shoe rack, then you can organize your shoes. If you buy a hat rack, then your hats won't be lying all around the house. You bought a shoe rack and your hats are lying all around the house.

A. You can organize your shoes and you didn't buy a hat rack.
B. You can organize your shoes and your hats aren't lying all over the house.
C. You can't organize your shoes and didn't buy a hat rack.
D. None of the above is warranted.

24. Study the information given below. If a logical conclusion is given, select that conclusion. If none of the conclusions given is warranted, select the option expressing this condition.

If I am your sister's son, then I am your nephew. I am not your sister's son.

A. I am your nephew.
B. I am not your nephew.
C. If I am your nephew, then I am your sister's son.
D. None of the above is warranted.

CLAST Exercises for Section 3.7 of *Thinking Mathematically*

CLAST SKILL II.E.3 *The student will draw logical conclusions from data.*

See page 148 of this book to learn about this CLAST skill.

1. Given that: i.   All mice live in holes;
                   ii.   All rodents live in holes;
determine which conclusion can be logically deduced.

A. All mice are rodents.   B. No mice are rodents.
C. All rodents are mice.   D. None of the above.

2. Given that: i.   All elephants are huge;
                   ii.   No avians are huge;
determine which conclusion can be logically deduced.

A. All elephants are avians.   B. Some elephants are avians.
C. No elephants are avians.   D. None of the above.

3. Given that: i.   All Greek epic poetry is classic;
                   ii.   *The Aeneid* is not Greek epic poetry;
determine which conclusion can be logically deduced.

A. *The Aeneid* is classic.   B. *The Aeneid* is too violent.
C. *The Aeneid* is not classic.   D. None of the above.

CLAST SKILL II.E.4 *The student will recognize that an argument may not be valid even though its conclusion is true.*

See page 151 of this book to learn about this CLAST skill.

4. All of the arguments A - D have true conclusions, but one of the arguments is not valid. Select the argument that is **not** valid.

A. All cats are hunters and some cats eat mice. Therefore, some hunters eat mice.
B. All clocks are timepieces and some clocks aren't accurate. Thus, some timepieces aren't accurate.
C. All diamonds are gemstones and some diamonds are worthless. Therefore, some gemstones are worthless.
D. All coins are currency and some currency is silver. Therefore, some coins are silver.

5. All of the arguments A - D have true conclusions, but one of the arguments is not valid. Select the argument that is **not** valid.

A. All subatomic particles are miniscule and all protons are subatomic particles. Therefore, all protons are miniscule.

B. All students pay fees and some students take loans. Therefore, some people who pay fees take loans.

C. All police officers are public servants and some public servants wear uniforms. Therefore, some police officers wear uniforms.

D. If one follows elephants with a shovel, then one works in a circus. Opticians don't work in a circus. Therefore, opticians don't follow elephants with shovels.

6. All of the arguments A - D have true conclusions, but one of the arguments is not valid. Select the argument that is **not** valid.

A. Some athletes are scholars. Some scholars win awards. Therefore, some athletes win awards.

B. All bears are dangerous animals. Some bears are brown. Therefore, some dangerous animals are brown.

C. All poets are writers. Some creative people aren't writers. Therefore, some creative people aren't poets.

D. All poodles are dogs. No cats are dogs. Therefore, no poodles are cats.

CLAST SKILL IV.E.1 *The student draws logical conclusions when the facts warrant them.*

See page 161 of this book to learn about this CLAST skill.

7. Study the information given below. If a logical conclusion is given, select that conclusion. If none of the conclusions given is warranted, select the option expressing this condition.

All snails move slowly.
Sluggo moves slowly.

A. Sluggo is a snail.   B. Sluggo isn't a snail.
C. Some snails don't move slowly.   D. None of the above is warranted.

SECTION 3.7

8. Study the information given below. If a logical conclusion is given, select that conclusion. If none of the conclusions given is warranted, select the option expressing this condition.

All big stars are flaming balls of gas.
All radio talk show hosts are flaming balls of gas.

A. No radio talk show hosts are big stars.   B. All radio talk show hosts are big stars.
C. All big stars are radio talk show hosts.   D. None of the above is warranted.

9. Study the information given below. If a logical conclusion is given, select that conclusion. If none of the conclusions given is warranted, select the option expressing this condition.

All libraries are quiet places.
No monkey houses are quiet places.
Sheila lives in a monkey house.

A. If one doesn't live in a library, then one lives in a monkey house.
B. All quiet places are libraries.
C. Sheila doesn't live in a library.
D. None of the above is warranted.

10. Study the information given below. If a logical conclusion is given, select that conclusion. If none of the conclusions given is warranted, select the option expressing this condition.

All of the people in my study group came to the study session.
All of the people who didn't bring calculators failed the test.
None of the people who came to the study session failed the test.
All of the people who solved the extra credit problem were in my study group.

A. All of the people who passed the test solved the extra credit problem.
B. All of the people who solved the extra credit problem failed the test.
C. None of the people who didn't bring calculators solved the extra credit problem.
D. None of the above is warranted.

CLAST Exercises for Section 4.1 of *Thinking Mathematically*.

CLAST SKILL II.A.2 *The student will recognize the role of the base number in determining place value in the base-ten numeration system.*

See page 11 of this book to learn about this CLAST skill.

1. Select the place value associated with the underlined digit. $\qquad$ 43$\underline{0}$1.25

A. 10^2 B. 10   C. 0   D. 10^0

2. Select the correct expanded notation for 200.04.

A. $\left(2 \times 10^3\right) + \left(4 \times \dfrac{1}{10^2}\right)$ B. $\left(2 \times 10^3\right) + \left(4 \times \dfrac{1}{10^3}\right)$

C. $\left(2 \times 10^2\right) + \left(4 \times \dfrac{1}{10}\right)$ D. $\left(2 \times 10^2\right) + \left(4 \times \dfrac{1}{10^2}\right)$

3. Select the numeral for $\left(8 \times 10^3\right) + \left(2 \times 10^1\right) + \left(9 \times \dfrac{1}{10^3}\right)$

A. 82.9   B. 829   C. 8020.009   D. 802.9

4. Select the correct expanded notation for 4601.3.

A. $\left(4 \times 10^4\right) + \left(6 \times 10^3\right) + \left(1 \times 10^1\right) + \left(3 \times \dfrac{1}{10^2}\right)$

B. $\left(4 \times 10^3\right) + \left(6 \times 10^2\right) + \left(1 \times 10^1\right) + \left(3 \times \dfrac{1}{10}\right)$

C. $\left(4 \times 10^4\right) + \left(6 \times 10^3\right) + \left(1 \times 10^1\right) + \left(3 \times \dfrac{1}{10}\right)$

D. $\left(4 \times 10^3\right) + \left(6 \times 10^2\right) + \left(1 \times 10^0\right) + \left(3 \times \dfrac{1}{10}\right)$

5. Select the place value associated with the underlined digit.   993.8<u>7</u>

A. 10^2 B. 10^1 C. $\dfrac{1}{10^2}$ D. $\dfrac{1}{10}$

6. Select the numeral for $\left(3 \times 10^2\right) + \left(6 \times 10^0\right) + \left(2 \times \dfrac{1}{10^2}\right)$

A. 306.02   B. 36.002   C. 306.002   D. 36.02

7. Select the place value associated with the underlined digit.   400<u>8</u>.092

A. 10^1 B. 10^8 C. 10^0 D. $\dfrac{1}{10^1}$

8. Select the correct expanded notation for 505.05

A. $\left(5 \times 10^2\right) + \left(5 \times 10^0\right) + \left(5 \times \dfrac{1}{10^2}\right)$ B. $\left(5 \times 10^2\right) + \left(5 \times 10^0\right) + \left(5 \times \dfrac{1}{10^1}\right)$

C. $\left(5 \times 10^3\right) + \left(5 \times 10^1\right) + \left(5 \times \dfrac{1}{10^1}\right)$ D. $\left(5 \times 10^3\right) + \left(5 \times 10^1\right) + \left(5 \times \dfrac{1}{10^2}\right)$

9. Select the numeral for $\left(7 \times 10^8\right) + \left(4 \times 10^3\right) + \left(5 \times \dfrac{1}{10^5}\right)$

A. 70,000,400.000 005   B. 700,004,000.000 05

C. 700,000,400.000 05   D. 70,004,000.000 005

CLAST Exercises for Section 5.1 of *Thinking Mathematically.*

CLAST SKILL IV.A.3 *The student solves problems that involve the structure and logic of arithmetic.*

See page 27 of this book to learn about this CLAST skill.

1. How many positive multiples of 4 are divisors of 84?
A. 1   B. 2   C. 3   D. 12

2. Find the largest divisor of 96 that is also a divisor of 52.
A. 13   B. 4   C. 2   D. 8

3. How many positive multiples of 5 leave a remainder of 2 when divided into 77?
A. 0   B. 25   C. 15   D. 3

4. Find the smallest multiple of 12 that is also a multiple of 16.
A. 96   B. 32   C. 48   D. 64

5. How many positive integers leave a remainder of 4 when divided into 74 and leave a remainder of 8 when divided into 85?
A. 0   B. 1   C. 7   D. 3

6. How many positive divisors of 36 are not divisors of 48?
A. 2   B. 1   C. 3   D. 0

7. Find the smallest positive number that leaves a remainder of 6 when divided into 50 and is not a prime number.
A. 11   B. 22   C. 4   D. 9

8. Find the largest divisor of 72 that is also a divisor of 60.
A. 4   B. 18   C. 9   D. 12

9. How many positive integers leave a remainder of 3 when divided into 48 and leave a remainder of 4 when divided into 44?
A. 1   B. 5   C. 4   D. 0

10. Find the smallest multiple of 12 that is also a multiple of 15.
A. 120   B. 30   C. 180   D. 60

SECTION 5.2

CLAST SKILL I.C.2 *The student will apply the order of operations agreement to computations involving numbers and variables.*

See page 66 of this book to learn about this CLAST skill.

1. $(6-4)^3 \div 2 + 2 =$

A. 6   B. 2

C. 78   D. 38

2. $20 - 16 \times 2 \div 4 =$

A. 2   B. 3

C. 12   D. -3

3. $8 + 4^2 \div (3^2 - 1) =$

A. 12   B. 10

C. 3   D. 17

4. $18 + 2 \times 5 - 4 =$

A. 96   B. 80

C. 20   D. 24

5. $12 \times 5^2 \div (14 - 10 + 2) =$

A. 50   B. 150

C. 600   D. 1800

CLAST Exercises for Section 5.2 of *Thinking Mathematically*.

CLAST SKILL II.A.1 *The student will recognize the meaning of exponents.*

See page 9 of this book to learn about this CLAST skill.

6. $(4^3)(3^2) =$

A. $(4 + 4 + 4)(3 + 3)$
B. $(3 \times 3 \times 3 \times 3)(2 \times 2 \times 2)$

C. $(4 \times 3)(3 \times 2)$
D. $(4 \times 4 \times 4)(3 \times 3)$

7. $5(2^4) =$

A. $(5^4)(2)$
B. $5(2 \times 2 \times 2 \times 2)$

C. $10 \times 10 \times 10 \times 10$
D. $10 + 10 + 10 + 10$

8. $10^2 + 5^2 =$

A. $(10 + 5)^2$
B. $(10)(10) + (5)(5)$

C. $(10 + 5)^4$
D. $(10)(10) + (5)(2)$

9. $(3^2)^3 =$

A. 3^5
B. 3^9

C. $(3 \times 2)^3$
D. $\left(3^2 \times 3^2 \times 3^2\right)$

10. $6^3 - 5^3 =$

A. $(6 - 5)^3$
B. $(6 - 5)^0$

C. $6 \times 6 \times 6 - 5 \times 5 \times 5$
D. $3(6 - 5)$

CLAST Exercises for Section 5.3 of *Thinking Mathematically*.

CLAST SKILL I.A.1a *The student will add and subtract rational numbers.*

See page 1 of this book to learn about this CLAST skill.

1. $2\frac{1}{4} + \frac{3}{8} =$ A. $2\frac{1}{3}$ B. $\frac{7}{8}$ C. $2\frac{5}{8}$ D. $3\frac{1}{8}$

2. $-3 + \frac{2}{5} =$ A. $-3\frac{2}{5}$ B. $-1\frac{1}{5}$ C. $2\frac{3}{5}$ D. $-2\frac{3}{5}$

3. $5 - 3\frac{2}{7} =$ A. $1\frac{5}{7}$ B. $2\frac{2}{7}$ C. $4\frac{1}{7}$ D. $1\frac{2}{7}$

4. $-\frac{1}{4} - 3 =$ A. $-2\frac{3}{4}$ B. $2\frac{3}{4}$ C. $3\frac{1}{4}$ D. $-3\frac{1}{4}$

5. $-2 + 1\frac{1}{3} =$ A. $3\frac{1}{3}$ B. $-\frac{2}{3}$ C. $-1\frac{1}{3}$ D. $\frac{2}{3}$

CLAST SKILL I.A.1b *The student will multiply and divide rational numbers.*

See page 2 of this book to learn about this CLAST skill.

6. $\frac{1}{4} \times 1\frac{1}{3} =$ A. $1\frac{1}{12}$ B. $\frac{1}{3}$ C. $2\frac{5}{8}$ D. 4

7. $\left(-\frac{2}{3}\right) \div \left(-\frac{1}{5}\right) =$ A. $-3\frac{1}{3}$ B. $\frac{2}{15}$ C. $3\frac{1}{3}$ D. $\frac{3}{10}$

8. $6 \div 3\frac{1}{4} =$ A. $1\frac{11}{13}$ B. 8 C. $\frac{1}{2}$ D. $19\frac{1}{2}$

9. $(-4) \times 5\frac{1}{2} =$ A. $-\frac{1}{2}$ B. -10 C. -22 D. 22

10. $2\frac{1}{2} \div \frac{3}{4} =$ A. $\frac{4}{3}$ B. $\frac{3}{10}$ C. $5\frac{1}{3}$ D. $6\frac{2}{3}$

CLAST SKILL I.A.2a   *The student will add and subtract rational numbers in decimal form.*

See page 3 of this book to learn about this CLAST skill.

11. $1.78 + .235 =$ A. 2.015   B. 1.313   C. 4.13   D. 1.915

12. $14.82 - .864 =$ A. 13.218   B. 6.18   C. 13.956   D. 14.066

13. $-12.47 - 1.031 =$ A. 13.501   B. -13.501 C. -13.401 D. 13.401

14. $3.02 - 5.411 =$ A. -2.409 B. 2.409   C. -2.391 D. -2.211

15. $3.041 - 1.24 =$ A. 1.801   B. 2.17   C. 2.801   D. 2.917

CLAST SKILL I.A.2b   *The student will multiply and divide rational numbers in decimal form.*

See page 4 of this book to learn about this CLAST skill.

16. $2.51 \times 1.6 =$ A. .4016   B. 4.016   C. 3.106   D. 3.116

17. $6.55 \div .05 =$ A. .3275   B. 0076   C. .0131   D. 131

18. $-(.03) \times (-2.4) =$ A. $-.072$ B. .072   C. 7.2   D. -7.2

19. $(-.08) \div 1.6 =$ A. -20 B. $-.5$ C. .05   D. $-.05$

20. $7.3 \times (-6.8) =$ A. 44.384   B. −44.384   C. −49.64   D. −4.964

CLAST SKILL II.A.4 *The student will determine the order relation between real numbers.*

See page 16 of this book to learn about this CLAST skill.

21. Identify the symbol that should be placed in the blank to form a true statement.

$\dfrac{6}{11}$ _____ $-\dfrac{8}{9}$ A. =   B. <   C. >

22. Identify the symbol that should be placed in the blank to form a true statement.

3.75 _____ $\dfrac{15}{4}$ A. =   B. <   C. >

23. Identify the symbol that should be placed in the blank to form a true statement.

$1.\overline{3}$ _____ 1.33 A. =   B. <   C. >

24. Identify the symbol that should be placed in the blank to form a true statement.

$\dfrac{3}{7}$ _____ $\dfrac{7}{16}$ A. =   B. <   C. >

25. Identify the symbol that should be placed in the blank to form a true statement.

8.23 _____ 8.227 A. =   B. <   C. >

26. Identify the symbol that should be placed in the blank to form a true statement.

$\dfrac{15}{20}$ _____ $\dfrac{30}{40}$ A. =   B. <   C. >

CLAST SKILL IV.A.1 *The student solves real-world problems which do not require the use of variables and do not involve percent.*

See page 24 of this book to learn about this CLAST skill.

27. A house rents for $750 per month. In order to move into the house, you need to pay two months' rent, a security deposit of $500 and a utility deposit of $200. How much will it cost to move into the house?

A. $1250   B. $2,200   C. $1,440   D. $2,000

28. Ed estimates that his family's monthly expenses for transportation are $200 per vehicle for vehicle payments, $30 per vehicle for basic insurance, $60 per vehicle for gasoline and maintenance, and additional insurance costs of $20 per driver. Find his family's estimated monthly transportation expenses if the family owns two cars and one truck and has four drivers.

A. $660   B. $1,220   C. $950   D. $930

29. Marie's college instructors tell her that she should spend three hours studying on her own for each hour spent in class. Additionally, she works four hours per day for five days each week, and spends 3 hours per week traveling to and from work and school. Find the total amount of time she should spend each week on school, studies, work and related travel, assuming that she spends 15 hours per week in class.

A. 68 hours   B. 80 hours   C. 95 hours   D. 83 hours

30. To park in the campus parking garage costs $1.00 for the first hour, plus 75¢ for each additional hour. Fractional parts of hours are rounded up. The parking garage is closed from 10:00 p.m. to 7:00 a.m. During that time, the hourly rate does not apply, but there is an extra fee of $10.00 for cars left overnight, and the hourly rate again takes effect at 7:00 a.m. What is the total charge for parking if you arrive at 9:05 a.m. and leave at 11:15 a.m. the same day?

A. $3.25   B. $13.25   C. $2.50   D. $2.25

SECTION 5.3

31. The homeowners association for Anthony's housing development dictates that a special fee for road improvement can be assessed if at least two thirds of the homeowners approve. There are 42 homeowners in Anthony's housing development. Anthony has been trying to convince them to vote in favor of the road improvement assessment. He has received commitments from a number of homeowners already, and he finds that he needs commitments from 3 more homeowners in order to guarantee approval of the assessment. How many homeowners (including Anthony) have already committed to support the assessment?

A. 14   B. 28   C. 31   D. 25

32. For a mid-sized car, a car rental agency charges $15 per day, plus 20¢ per mile, and there is an optional collision damage waiver charge of $6.00 per day. Find the total cost to a person who rents the car for three days, travels a total of 400 miles, and does not purchase the optional collision damage waiver.

A. $125   B. $445   C. $143   D. $105

CLAST Exercises for Section 5.4 of *Thinking Mathematically*.

CLAST SKILL I.C.1a *The student will add and subtract real numbers.*

See page 64 of this book to learn about this CLAST skill.

1. $\sqrt{6} + \sqrt{54} =$ A. $10\sqrt{6}$ B. $4\sqrt{6}$ C. $2\sqrt{15}$ D. $\sqrt{15}$

2. $4\sqrt{2} + \sqrt{2} =$ A. $\sqrt{5}$ B. 16 C. $5\sqrt{2}$ D. $\sqrt{5}$

3. $\sqrt{27} - 2\sqrt{3} =$ A. $4\sqrt{6}$ B. $7\sqrt{3}$ C. $2\sqrt{3}$ D. $\sqrt{3}$

4. $\sqrt{2} - \sqrt{8} =$ A. $-\sqrt{2}$ B. $-\sqrt{6}$ C. $\sqrt{2}$ D. $\sqrt{6}$

5. $5 + \sqrt{20} - \sqrt{80} =$ A. $4\sqrt{5}$ B. $-\sqrt{5}$ C. $-10\sqrt{5}$ D. $5 - 2\sqrt{5}$

6. $\sqrt{49} - 7 =$ A. $\sqrt{42}$ B. $7\sqrt{7}$ C. 0 D. $-7\sqrt{7}$

7. $3 - 2\pi + 6\pi =$ A. $3 - 4\pi$ B. 5π C. 7π D. $3 + 4\pi$

8. $2\pi + 8 + 6\pi =$ A. 16π B. $8\pi + 8$ C. $10 + 6\pi$ D. $10\pi + 6$

CLAST SKILL I.C.1b *The student will multiply and divide real numbers.*

See page 65 of this book to learn about this CLAST skill.

9. $\sqrt{2} \times \sqrt{10} =$ A. 40 B. 20 C. $2\sqrt{5}$ D. $4\sqrt{5}$

10. $\sqrt{18} \times \sqrt{3} =$ A. $3\sqrt{6}$ B. 54 C. 18 D. $9\sqrt{3}$

11. $\dfrac{5}{\sqrt{2}} =$ A. $\dfrac{\sqrt{5}}{2}$ B. $\dfrac{5\sqrt{2}}{4}$ C. $\dfrac{5\sqrt{2}}{2}$ D. $5\sqrt{2}$

12. $\dfrac{8}{\sqrt{6}} =$ A. $\dfrac{\sqrt{6}}{48}$ B. $\dfrac{4\sqrt{6}}{3}$ C. $\dfrac{\sqrt{6}}{6}$ D. $\dfrac{2\sqrt{6}}{9}$

13. $\sqrt{10} \times \sqrt{6} =$ A. $2\sqrt{15}$ B. $4\sqrt{15}$ C. 60 D. 3600

14. $\dfrac{1}{\sqrt{3}} =$ A. $\dfrac{\sqrt{3}}{9}$ B. $\dfrac{1}{9}$ C. $\dfrac{1}{3}$ D. $\dfrac{\sqrt{3}}{3}$

15. $\sqrt{5} \times \sqrt{7} =$ A. 35 B. $\sqrt{35}$ C. $2\sqrt{3}$ D. 12

16. $\dfrac{4}{\sqrt{5}} =$ A. $4\sqrt{5}$ B. $\dfrac{4}{25}$ C. $\dfrac{4\sqrt{5}}{25}$ D. $\dfrac{4\sqrt{5}}{5}$

CLAST SKILL II.A.4 *The student will determine the order relation between real numbers.*

See page 16 of this book to learn about this CLAST skill.

17. Identify the symbol that should be placed in the blank to form a true statement.

6.14 ____ $\sqrt{35}$ A. $=$ B. $<$ C. $>$

18. Identify the symbol that should be placed in the blank to form a true statement.

$\sqrt{80}$ ____ $9.1\overline{3}$ A. $=$ B. $<$ C. $>$

19. Identify the symbol that should be placed in the blank to form a true statement.

$\dfrac{\sqrt{9}}{2}$ ____ 1.50000 A. $=$ B. $<$ C. $>$

20. Identify the symbol that should be placed in the blank to form a true statement.

$-\sqrt{10}$ ____ $-\sqrt{8}$ A. $=$ B. $<$ C. $>$

21. Identify the symbol that should be placed in the blank to form a true statement.

$\sqrt{6}$ ____ 3.14 A. $=$ B. $<$ C. $>$

CLAST Exercises for Section 5.6 of *Thinking Mathematically.*

CLAST SKILL I.C.3 *The student will use scientific notation in calculations involving very large or very small measurements.*

See page 68 of this book to learn about this CLAST skill.

1. $.00064 \div 1,600,000 =$
   A. 4.00×10^{11} B. 4.00×10^{10} C. 4.00×10^{-10} D. 4.00×10^{2}

2. $(2.1 \times 10^{4}) \times (3.0 \times 10^{-6}) =$
   A. $.063$ B. $-.063$ C. 630 D. -6300

3. $2,100,000,000 \div .00007 =$
   A. 3.00×10^{23} B. 3.00×10^{-23} C. 3.00×10^{-13} D. 3.00×10^{13}

4. $(3.8 \times 10^{-8}) \times (2.0 \times 10^{12}) =$
   A. 5.8×10^{-96} B. 7.6×10^{4} C. 5.8×10^{-4} D. 7.6×10^{-20}

5. $.00045 \div .00009 =$
   A. 5.00×10^{-8} B. 4.15×10^{-8} C. 5 D. 4.15×10^{8}

6. $(7.5 \times 10^{5}) \div (3.0 \times 10^{8}) =$
   A. 2500 B. $-.0025$ C. $.0025$ D. $-.025$

7. $(4.0 \times 10^{-8}) \times (8.1 \times 10^{9}) =$
   A. 32.4 B. $.324$ C. $-.324$ D. 324

8. $.0012 \div 6,000,000 =$
   A. 2.00×10^{-10} B. 2.00×10^{3} C. 2.00×10^{-9}D. 2.00×10^{9}

CLAST Exercises for Section 5.7 of *Thinking Mathematically*

CLAST SKILL III.A.1 *The student infers relations between numbers in general by examining particular number pairs.*

See page 21 of this book to learn about this CLAST skill.

1. Identify the missing term in the following arithmetic progression.

10, 6, 2, –2, –6, _____

A. –8   B. –10   C. 12   D. –4

2. Find the missing term in the following geometric progression:

–6, –12, –24, –48, _____

A. –72   B. 72   C. 96   D. –96

3. Find the missing term in the following harmonic progression:
$$\frac{1}{2}, \frac{1}{5}, \frac{1}{8}, \frac{1}{11}, \frac{1}{14}, \text{———}$$

A. $\frac{1}{25}$ B. $\frac{1}{151}$ C. $\frac{1}{17}$ D. $\frac{1}{19}$

4. Find the missing term in the following geometric progression:

$$-9, 6, -4, \frac{8}{3}, -\frac{16}{9} \text{———}$$

A. $\frac{48}{27}$ B. $\frac{24}{12}$ C. $\frac{144}{27}$ D. $-\frac{48}{27}$

5. Identify the missing term in the following arithmetic progression.

–8, –5, –2, 1, 4 _____

A. 3   B. 1   C. 7   D. 5

6. Find the missing term in the following harmonic progression:

$$\frac{1}{3}, \frac{1}{10}, \frac{1}{17}, \frac{1}{24}, \frac{1}{31}, \underline{\quad}$$

A. $\frac{1}{38}$ B. $\frac{1}{7}$ C. $\frac{1}{51}$ D. $\frac{1}{61}$

7. Find the missing term in the following geometric progression:

$$10, 4, \frac{8}{5}, \frac{16}{25}, \frac{32}{125}, \underline{\quad}$$

A. $\frac{48}{150}$ B. $\frac{64}{625}$ C. $\frac{5}{2}$ D. $\frac{160}{250}$

8. Find the missing term in the following harmonic progression:

$$\frac{1}{8}, \frac{1}{13}, \frac{1}{18}, \frac{1}{23}, \frac{1}{28}, \underline{\quad}$$

A. $\frac{2}{51}$ B. $\frac{1}{33}$ C. $\frac{1}{5}$ D. $\frac{1}{51}$

9. Find the missing term in the following geometric progression:

$-2, 6, -18, 54, \underline{\quad}$

A. -18 B. 72 C. 162 D. -162

10. Find the missing term in the following geometric progression:
$$6, -\frac{9}{2}, \frac{27}{8}, -\frac{81}{32}, \frac{243}{128}, \underline{\quad}$$

A. $-\frac{324}{160}$ B. $-\frac{729}{512}$ C. $\frac{729}{512}$ D. $-\frac{936}{384}$

CLAST Exercises for Section 6.1 of *Thinking Mathematically.*

CLAST SKILL I.C.2 *The student will apply the order of operations agreement to computations involving numbers and variables.*

See page 66 of this book to learn about this CLAST skill.

1. $\dfrac{1}{4} \times \dfrac{1}{2} - \dfrac{1}{2} + \dfrac{1}{4} =$

   A. $-\dfrac{1}{4}$ B. $-\dfrac{1}{8}$ C. 0 D. $\dfrac{1}{16}$

2. $3 \times (2r)^2 - r^2 =$

   A. $5r^2$ B. $35r^2$ C. $3 + 3r^2$ D. $11r^2$

3. $\dfrac{1}{3} - \dfrac{2}{3} \times \dfrac{4}{3} \div \dfrac{2}{5} =$

   A. $-\dfrac{25}{18}$ B. $-\dfrac{10}{9}$ C. $-\dfrac{17}{9}$ D. $-\dfrac{1}{45}$

4. $x + \dfrac{2x^2}{x} =$

   A. $3x$ B. $5x^2$ C. $2x^2$ D. $3x^2$

5. $\dfrac{2}{3} - 4\left(\dfrac{1}{4} - \dfrac{1}{2}\right) =$

   A. $\dfrac{5}{6}$ B. $-\dfrac{5}{6}$ C. $-\dfrac{1}{3}$ D. $\dfrac{5}{3}$

6. $2(s + 5) - 3 =$

   A. $2s + 2$ B. $10s - 3$ C. $7s$ D. $2s + 7$

7. $\dfrac{4(3x-4y)}{4} - (x-3y) =$

A. $2x - y$ B. $5x - 7y$ C. $2x - 3y$ D. $2x - 7y$

8. $4 + t^2 - 4(3t)^2 =$

A. $4 - 11t^2$ B. $4 - 35t^2$ C. $-7t^2$ D. $-32t^2$

CLAST SKILL I.C.5 *The student will use given formulas to compute results when geometric measurements are not involved.*

See page 73 of this book to learn about this CLAST skill.

9. Given $r = 3(s+t)^2 + s$, find r if s = 4 and t = 1.

A. 29   B. 79   C. 172   D. 76

10. Given a = N(2 + L) find a if L = 4 and N = 10.

A. 48   B. 40   C. 60   D. 42

11. The formula $C = \dfrac{5}{9}(F - 32°)$ is used for converting temperature from Fahrenheit (*F*) to Celsius (*C*). What is the temperature on the Celsius scale when the Fahrenheit temperature is 149°?

A. 65°   B. 50.8°   C. –15.4°   D. 100.6°

12. If a rock is dropped from a 300-foot high bridge, the formula $D = 300 - 16T^2$ gives the rock's distance from the ground (in feet) after it has fallen T seconds. Find the rock's distance from the ground after 3 seconds.

A. 219 feet   B. 252 feet   C. 144 feet   D. 156 feet

13. The formula $S = \dfrac{1}{4}(Q + M + 2E)$ gives the total score in a course, where Q is the student's quiz average, M is the midterm exam score, and E is the final exam score. Find Marcia's total score if her quiz average is 82, her midterm score is 76, and her final exam score is 90.

A. 81   B. 87.3   C. 84.5   D. 62

SECTION 6.1

CLAST SKILL II.C.1 *The student will use properties of operations correctly.*

See page 83 of this book to learn about this CLAST skill.

14. Choose the expression equivalent to the following: $5(x + y)$

A. $5 + (x + y)$ B. $5x + 5y$ C. $5x + y$ D. $5xy$

15. Choose the expression equivalent to the following: $9s + 5t$

A. $9t + 5s$ B. $5t + 9s$ C. $9(s + 5t)$ D. $9s(5 + t)$

16. Choose the expression equivalent to the following: $4a + 2b$

A. $2(a + b)$ B. $2(2a + b)$ C. $2a + 4b$ D. $-4a - 2b$

17. Choose the expression equivalent to the following: $-(3x) + 3x$

A. $3(-x)$ B. $-6x$ C. $-9x^2$ D. 0

18. Choose the expression equivalent to the following: $6x(y - 2)$

A. $-12xy$ B. $4x - y$ C. $(y - 2)6x$ D. $6x(2 - y)$

19. Choose the statement that is <u>not</u> true for all real numbers.

A. $2(x) + 2(y) = 2(x + y)$ B. $(x + y)(x - y) = (x - y)(x + y)$

C. $x + \dfrac{1}{x} = 1$ D. $(3 + x) + y = 3 + (x + y)$

20. Choose the statement that is <u>not</u> true for all real numbers.

A. $(5x + 8y) = (5y + 8x)$ B. $2(xy) = (2x)y$

C. $-(xy) + xy = 0$ D. $x(y + z) = xy + xz$

CLAST Exercises for Section 6.2 of *Thinking Mathematically*.

CLAST SKILL III.C.2 *The student uses applicable properties to select equivalent equations and inequalities.*

See page 96 of this book to learn about this CLAST skill.

1. Choose the equation equivalent to the following:

$4x = 2 - 5x$

A. $4 = 2x - 5$ B. $9x = 2$ C. $0 = 2 - x$ D. $x = \dfrac{2}{4} - 5x$

2. Choose the equation equivalent to the following:

$x + 8 = 4x - 3$

A. $x = 4x - 11$ B. $x = \dfrac{1}{2}x - \dfrac{3}{8}$ C. $8 = 5x - 3$ D. $x - 5 = 4x$

3. Choose the equation equivalent to the following:

$10 - 6x = 4x - 12$

A. $10 = 4x - 18$ B. $10 = 4x - 6$ C. $10 - 10x = -8$ D. $\dfrac{10}{4} - \dfrac{6}{4}x = x - 3$

4. Choose the equation equivalent to the following:

$\dfrac{1}{2}x + 1 = 4$

A. $x + 1 = 8$ B. $x + 2 = 8$ C. $x + \dfrac{3}{2} = \dfrac{9}{2}$ D. $x + \dfrac{1}{2} = 2$

5. Choose the equation equivalent to the following:

$3x + 1 = 2x - 1$

A. $3x - 1 = 2x + 1$ B. $3x = 2x$ C. $5x + 1 = -1$ D. $2x - 1 = 3x + 1$

SECTION 6.2

CLAST SKILL I.C.4a *The student will solve linear equations.*

See page 70 of this book to learn about this CLAST skill.

6. If $4x - 3 = 5$, then   A. $x = \dfrac{1}{2}$ B. $x = 2$ C. $x = 8$ D. $x = -2$

7. If $7 - 4y = 2$, then   A. $y = \dfrac{9}{4}$ B. $y = -\dfrac{9}{4}$ C. $y = -\dfrac{5}{4}$ D. $y = \dfrac{5}{4}$

8. If $-10 = 3(q - 2)$, then

A. $q = -11$ B. $q = -15$ C. $q = -\dfrac{4}{3}$ D. $q = -\dfrac{8}{3}$

9. If $5(2x + 3) = 4(x - 1)$, then

A. $x = -\dfrac{19}{6}$ B. $x = -\dfrac{19}{14}$ C. $x = -\dfrac{6}{7}$ D. $x = \dfrac{6}{7}$

10. If $3r - 9 = 5(2r + 3)$, then

A. $r = -11$ B. $r = 17$ C. $r = -\dfrac{12}{7}$ D. $r = -\dfrac{24}{7}$

11. If $9a + 12 = 5a$, then

A. $a = 3$ B. $a = -3$ C. $a = \dfrac{6}{7}$ D. $a = \dfrac{7}{6}$

12. If $3(x + 2) - 2 = x - 6$, then

A. $x = \dfrac{10}{3}$ B. $x = -\dfrac{10}{3}$ C. $x = -5$ D. $x = -\dfrac{14}{3}$

13. If $10 - 2(3t + 1) = 4(t + 6) - 6$, then

A. $t = -1$ B. $t = 1$ C. $t = -5$ D. $t = 5$

14. If $4z - 3(z - 5) = 8z - 1$, then

A. $z = -2$ B. $z = 2$ C. $z = \dfrac{16}{7}$ D. $z = -\dfrac{4}{7}$

15. If $6x + 4(x + 1) = 9(x - 2)$, then

A. $x = -\dfrac{19}{22}$ B. $x = -19$ C. $x = 22$ D. $x = -22$

CLAST Exercises for Section 6.3 of *Thinking Mathematically.*

CLAST SKILL IV.C.2 *The student solves problems that involve the logic and structure of algebra.*

See page 101 of this book to learn about this CLAST skill.

1. Choose the equation that is equivalent to the verbal description.

The product of 4 and a number, *n*, is the number increased by 12.

A. $4n + 12 = 4$ B. $4(n + 12) = n$ C. $4n = n + 12$ D. $4n = 4(n + 12)$

2. Choose the equation that is equivalent to the verbal description.

When a number *x* is increased by 18 it is the same as decreasing twice the number by 4.

A. $18x = 4x - 2$ B. $x = 4(2x + 18)$ C. $x + 8x = 2(x - 4)$ D. $x + 18 = 2x - 4$

3. Choose the equation that is equivalent to the verbal description.

A number, *n*, increased by half of itself is the product of 2 and the number increased by 1.

A. $n + \dfrac{1}{2} = 2 + n + 1$ B. $n + \dfrac{1}{2}n = 2(n + 1)$

C. $\dfrac{1}{2}n = 2(n + 1)$ D. $n = \dfrac{1}{2} + 2(n + 1)$

SECTION 6.3

CLAST SKILL IV.C.1 *The student solves real-world problems involving the use of variables, aside from commonly-used geometric formulas.*

See page 98 of this book to learn about this CLAST skill.

4. Randall is raising a hog. He paid $85 for the hog, and estimates that it will cost $35 per month to feed and care for the hog. He will keep the hog until he has spent a total of $400, and then sell it. For how many months will he keep the hog?

A. 12   B. 11   C. 9   D. 10

5. Thelmira and Louisa are driving a beaten-up old Firebird that gets 12 miles per gallon. How far can they travel if they have a total of $48 to spend on gasoline, assuming that gasoline costs $1.60 per gallon?

A. 480 miles   B. 360 miles   C. 6.4 miles   D. 640 miles

6. Jerry earns $7.00 per hour working part-time. His employer deducts 12.5% of his gross earnings for federal income tax withholding, and 7.5% of his gross earnings for Social Security. If Jerry's net pay was $140, how many hours did he work?

A. 20   B. 17   C. 25   D. 47

7. The college Health Center has budgeted 225 labor hours for its Health Counseling program. Under the program, students are invited to come to the Health Center to receive free advice on healthy living and nutrition, along with health-related gifts. Based on past experience, they expect to spend one-half of a labor hour on each student who attends, and that 15% of the students who receive invitations will attend. How many invitations should they send if they expect to stay within budget?

A. 3,000

B. 16,875

C. 750

D. 1,125

CLAST Exercises for Section 6.4 of *Thinking Mathematically.*

CLAST SKILL II.C.3 *The student will recognize statements of proportionality and variation.*

See page 87 of this book to learn about this CLAST skill.

1. Three ditch-diggers can dig 5 ditches in two days. Let D represent the number of ditches that 4 ditch-diggers can dig in two days. Select the correct statement of the given condition.

A. $\dfrac{3}{5} = \dfrac{D}{4}$
B. $\dfrac{3}{5} = \dfrac{4}{D}$
C. $\dfrac{3}{5} = \dfrac{2}{D}$
D. $\dfrac{3}{5} = \dfrac{D}{2}$

2. George studies for 10 hours in order to prepare for three final exams. Let H be the number of hours he would study to prepare for 5 final exams. Select the correct statement of the given condition.

A. $\dfrac{10}{5} = \dfrac{3}{H}$
B. $10H = 15$
C. $\dfrac{H}{3} = \dfrac{5}{10}$
D. $\dfrac{H}{5} = \dfrac{10}{3}$

3. In a certain habitat, the number of lizards is inversely proportional to the number of cats. There are 120 lizards when there are 2 cats. Let L be the number of lizards when there are 4 cats. Select the correct statement of the given condition.

A. $\dfrac{120}{2} = \dfrac{L}{4}$
B. $\dfrac{120}{2} = \dfrac{4}{L}$
C. $\dfrac{120}{L} = \dfrac{4}{2}$
D. $8 = 120L$

4. At the Unified Parcel Shipping Company, the cost of insuring a package for a 1500 mile delivery varies directly as the value of the package. It costs \$8.00 to insure a \$500 package. Let C be the cost of insuring a \$700 package. Select the correct statement of the given condition.

A. $500C = 5600$
B. $5600C = 500$
C. $1500C = 700$
D. $500C = 12000$

5. For a certain theatrical production, the weekly attendance varies inversely as the number of weeks that the production has been running. After the production has been running 10 weeks, the weekly attendance is 2500. Let A be the weekly attendance after the production has been running for 15 weeks. Select the correct statement of the given condition.

A. $\dfrac{10}{2500} = \dfrac{15}{A}$
B. $\dfrac{10}{A} = \dfrac{15}{2500}$
C. $\dfrac{10}{A} = \dfrac{2500}{15}$
D. $\dfrac{15}{2500} = \dfrac{A}{10}$

6. For a bread recipe, the number of cups of water used varies directly as the number of cups of flour. When 10 cups of flour are used and the bread is baked at 350 degrees, the number of cups of water used is 3.25. Let W be the number of cups of water used when 14 cups of flour are used. Select the correct statement of the given condition.

A. $32.5W = (10)(350)$

B. $350W = \dfrac{10}{3.25}$

C. $\dfrac{W}{14} = \dfrac{3.25}{10}$

D. $\dfrac{W}{10} = \dfrac{3.25}{14}$

CLAST SKILL IV.C.1 *The student solves real-world problems involving the use of variables, aside from commonly-used geometric formulas.*

See page 98 of this book to learn about this CLAST skill.

7. At Cory's Catering, the cost of catering a retirement party varies directly as the number of people who will attend the party. If it costs $400 to cater a retirement party for 60 people in honor of Bill, who is retiring after 30 years, how much will it cost to cater of retirement party for 35 people for Jill, who is retiring after 15 years?

A. $200   B. $685.71   C. $342   D. $233.33

8. The amount of sugar included in a recipe for brownies varies directly as the number of servings being prepared. If one and a half cups of sugar are required for 12 servings, how many cups of sugar are required for 20 servings?

A. 1.6   D. 4   C. 2.5   D. 3.2

9. Suppose that for a certain make of automobile, the value of the automobile varies inversely as the age of the automobile. If such a car is worth $4,500 when it is 10 years old, find the value when it is 15 years old.

A. $6750   B. $675   C. $3,333.33   D. $3,000

10. A broker finds that the weekly demand for a certain precious metal varies inversely as the price of the metal. When the price is $5.00 per ounce, weekly demand is 40 ounces. Find the weekly demand when the price is $4.00 per ounce.

A. 24 ounces   B. 32 ounces   C. 50 ounces   D. 44 ounces

CLAST exercises for Section 6.5 of *Thinking Mathematically.*

CLAST SKILL III.C.2 *The student uses applicable properties to select equivalent equations and inequalities.*

See page 96 of this book to learn about this CLAST skill.

1. Choose the inequality equivalent to the following:

$5(x + 1) > -20$

A. $x + 1 > -25$ B. $x + 1 > -4$ C. $x + 1 < -4$ D. $x + 1 < -25$

2. Choose the inequality equivalent to the following:

$3x - 6 > -6$

A. $x - 2 > -2$ B. $x - 2 < -2$ C. $x - 2 > 2$ D. $x - 2 < 2$

3. Choose the inequality equivalent to the following:

$-12x > 3$

A. $x > -9$ B. $x < -15$ C. $x > -\dfrac{1}{4}$ D. $x < -\dfrac{1}{4}$

4. If $x < 0$, then $4x^2 < -4x$ is equivalent to which of the following?

A. $x^2 > x$ B. $x^2 < x$ C. $x > -1$ D. $x < -1$

5. If $x > 0$, then $xy > x^2 - 2x$ is equivalent to which of the following?

A. $y > x - 2$ B. $y < x - 2$ C. $x^2 - 2x + xy > 0$ D. $x^2 - 2x + xy < 0$

CLAST SKILL I.C.4b
The student will solve linear inequalities.

See page 71 of this book to learn about this CLAST skill.

6. If $x + 3 < -2$, then

A. $x < 1$ B. $x > 1$ C. $x > -5$ D. $x < -5$

SECTION 6.5

7. If $3y - 4 \geq 1$, then

A. $y \geq \dfrac{5}{3}$　　B. $y \leq \dfrac{5}{3}$　　C. $y \leq 1$　　D. $y \geq 1$

8. If $5x + 1 > 3x - 1$, then

A. $x > \dfrac{3}{5}$　　B. $x < \dfrac{5}{3}$　　C. $x > -1$　　D. $x < 1$

9. If $3(r - 2) \leq 4r + 1$, then

A. $r \geq 7$　　B. $r \geq -7$　　C. $r \leq -7$　　D. $r \leq 7$

10. If $5(x - 8) + 22 < 4x - 10$, then

A. $x < 2$　　B. $x < 8$　　C. $x < 24$　　D. $x < \dfrac{8}{9}$

11. If $3(5 - a) \geq 4(a + 1) - 6$, then

A. $a \leq -\dfrac{17}{7}$　　B. $a \geq -\dfrac{17}{7}$　　C. $a \leq \dfrac{17}{7}$　　D. $a \geq \dfrac{17}{7}$

12. If $z - 2(z + 3) > 3(5 - z) + 1$, then

A. $z > 11$　　B. $z > 5$　　C. $z > \dfrac{1}{2}$　　D. $z > \dfrac{11}{2}$

13. If $4(x + 1) - 5x < 2(x - 3)$, then

A. $x < \dfrac{4}{3}$　　B. $x > \dfrac{4}{3}$　　C. $x > \dfrac{10}{3}$　　D. $x < \dfrac{10}{3}$

14. If $2 - 3q \geq 0$ then

A. $q \geq 0$　　B. $q \geq -5$　　C. $q \geq \dfrac{2}{3}$　　D. $q \leq \dfrac{2}{3}$

15. If $-(3t + 1) + 8 > 2(t + 1) - 1$, then

A. $t < \dfrac{6}{5}$　　B. $t < \dfrac{4}{5}$　　C. $t > \dfrac{4}{5}$　　D. $t > -\dfrac{4}{5}$

CLAST Exercises for Section 6.6 of *Thinking Mathematically.*

CLAST SKILL I.C.7 *The student will factor a quadratic expression.*

See page 75 of this book to learn about this CLAST skill.

1. Which is a linear factor of the following expression?
$10x^2 + 13x - 3$

A. $5x + 1$ B. $10x + 1$ C. $5x - 1$ D. $x + 3$

2. Which is a linear factor of the following expression?
$20x^2 + 7x - 6$

A. $5x - 2$ B. $5x + 3$ C. $4x - 3$ D. $5x + 6$

3. Which is a linear factor of the following expression?
$3x^2 - 5x - 12$

A. $3x - 2$ B. $3x + 1$ C. $x - 3$ D. $x + 12$

4. Which is a linear factor of the following expression?
$6x^2 + 7x - 20$

A. $6x - 5$ B. $2x + 5$ C. $6x + 5$ D. $3x - 10$

5. Which is a linear factor of the following expression?
$3x^2 - 5x - 2$

A. $3x - 2$ B. $x + 2$ C. $x + 1$ D. $3x + 1$

6. Which is a linear factor of the following expression?
$2x^2 + 9x + 4$

A. $2x + 9$ B. $x + 4$ C. $x + 2$ D. $2x - 1$

7. Which is a linear factor of the following expression?
$4x^2 + 19x - 5$

A. $2x + 5$ B. $2x - 5$ C. $x + 5$ D. $x - 5$

SECTION 6.6

8. Which is a linear factor of the following expression?
$4x^2+16x+15$

A. $4x+3$ B. $4x+15$ C. $2x+15$ D. $2x+5$

9. Which is a linear factor of the following expression?
$6x^2-5x+1$

A. $3x+1$ B. $6x+1$ C. $6x-1$ D. $2x-1$

10. Which is a linear factor of the following expression?
$8x^2-10x-3$

A. $8x+1$ B. $8x-1$ C. $4x+1$ D. $4x-1$

CLAST SKILL I.C.8 *The student will find the roots of a quadratic equation.*

See page 76 of this book to learn about this CLAST skill.

11. Find the correct solutions to this equation: $8x^2 - 2x - 1 = 0$

   A. $-\dfrac{1}{2}$ and $\dfrac{1}{4}$ B. 4 and -2

   C. $\dfrac{1}{2}$ and $-\dfrac{1}{4}$ D. $\dfrac{-1+\sqrt{7}}{8}$ and $\dfrac{-1-\sqrt{7}}{8}$

12. Find the correct solutions to this equation: $4x^2 + x - 1 = 0$

   A. $\dfrac{1+\sqrt{15}}{8}$ and $\dfrac{1-\sqrt{15}}{8}$ B. $\dfrac{1+\sqrt{17}}{8}$ and $\dfrac{1-\sqrt{17}}{8}$

   C. $\dfrac{1+\sqrt{15}}{2}$ and $\dfrac{1-\sqrt{15}}{2}$ D. $\dfrac{-1+\sqrt{17}}{8}$ and $\dfrac{-1-\sqrt{17}}{8}$

13. Find the correct solutions to this equation: $x^2 + 4x - 2 = 0$

   A. $-2+\sqrt{2}$ and $-2-\sqrt{2}$ B. $2+\sqrt{6}$ and $2-\sqrt{6}$

   C. $-2+\sqrt{6}$ and $-2-\sqrt{6}$ D. $2+\sqrt{2}$ and $2-\sqrt{2}$

14. Find the correct solutions to this equation: $4x^2 - 4x = 15$

   A. $\dfrac{5}{2}$ and $-\dfrac{3}{2}$ B. $-\dfrac{5}{2}$ and $\dfrac{3}{2}$

   C. $\dfrac{1+\sqrt{14}}{2}$ and $\dfrac{1-\sqrt{14}}{2}$ D. $\dfrac{1+4\sqrt{14}}{8}$ and $\dfrac{1-4\sqrt{14}}{8}$

15. Find the correct solutions to $3x^2 - 2 = 2x$

   A. $\dfrac{1+\sqrt{5}}{3}$ and $\dfrac{1-\sqrt{5}}{3}$ B. $\dfrac{-1+\sqrt{5}}{3}$ and $\dfrac{-1-\sqrt{5}}{3}$

   C. $\dfrac{1+\sqrt{7}}{3}$ and $\dfrac{1-\sqrt{7}}{3}$ D. $\dfrac{1+\sqrt{3}}{3}$ and $\dfrac{1-\sqrt{3}}{3}$

16. Find the correct solutions to: $x^2 + 6x + 2 = 0$

A. $-6 + \sqrt{7}$ and $-6 - \sqrt{7}$

B. $3 + 2\sqrt{7}$ and $3 - 2\sqrt{7}$

C. $-3 + \sqrt{11}$ and $-3 - \sqrt{11}$

D. $-3 + \sqrt{7}$ and $-3 - \sqrt{7}$

17. Find the correct solutions to: $x^2 = 8x - 4$

A. $4 + \sqrt{3}$ and $4 - \sqrt{3}$

B. $4 + 2\sqrt{3}$ and $4 - 2\sqrt{3}$

C. $8 + 2\sqrt{3}$ and $8 - 2\sqrt{3}$

D. $4 + 4\sqrt{3}$ and $4 - 4\sqrt{3}$

18. Find the correct solutions to: $12x^2 + 2x = 5 - 2x$

A. $\dfrac{\sqrt{15}}{6}$ and $-\dfrac{\sqrt{15}}{6}$

B. $-\dfrac{5}{6}$ and $\dfrac{1}{2}$

C. 10 and -6

D. -10 and 6

19. Find the correct solutions to: $4x^2 + x = x + 3$

A. $\dfrac{1 + \sqrt{3}}{4}$ and $\dfrac{1 - \sqrt{3}}{4}$

B. $\dfrac{-1 + \sqrt{3}}{4}$ and $\dfrac{-1 - \sqrt{3}}{4}$

C. 4 and -3

D. $\dfrac{\sqrt{3}}{2}$ and $-\dfrac{\sqrt{3}}{2}$

20. Find the correct solutions to: $-9x^2 + 15x - 4 = 0$

A. $\dfrac{4}{3}$ and $\dfrac{1}{3}$

B. $\dfrac{5}{6}$ and $-\dfrac{5}{6}$

C. $\dfrac{5 + 3\sqrt{41}}{6}$ and $\dfrac{5 - 3\sqrt{41}}{6}$

D. $\dfrac{3\sqrt{5}}{2}$ and $-\dfrac{3\sqrt{5}}{2}$

CLAST SKILL II.C.2 *The student will determine whether a particular number is among the solutions of a given equation or inequality.*

See page 85 of this book to learn about this CLAST skill.

21. For each of the statements given below, determine whether $x = -2$ is a solution.

    i. $(x + 1)(x + 2) = 0$
    ii. $|x + 2| = 0$
    iii. $x^2 + 12x + 20 \leq 0$

A. i, ii, and iii   B. i and ii only   C. i and iii only   D. ii and iii only

22. For each of the statements given below, determine whether $x = 5$ is a solution.

    i. $x^2 - 25 = 0$
    ii. $\left(x - \dfrac{1}{2}\right)(3x - 15) = 0$
    iii. $(x - 2)(x + 1) \leq 0$

A. i and ii only   B. i only   C. ii and iii only   D. i, ii, and iii

23. For each of the statements given below, determine whether $x = -3$ is a solution.

    i. $|x| < 2$
    ii. $2x + 1 = 3x - 4$
    iii. $2x^2 + 7x + 5 = 2$

A. i only   B. ii only   C. iii only   D. i, ii, and iii

24. For each of the statements given below, determine whether $x = \dfrac{1}{4}$ is a solution.

    i. $4x - 1 < x - 2$
    ii. $x + 1 = 2 - 3x$
    iii. $(x - 1)(2x - 1) = -3/8$

A. i only   B. ii only   C. iii only   D. ii and iii only

SECTION 6.6

25. For each of the statements given below, determine whether $x = -1$ is a solution.

    i. $\dfrac{1}{3}(5x - 1) = 2$

    ii. $2 + x \leq 2 - x$

    iii. $3x^2 + 2x = 1$

A. i and ii only   B. ii and iii only   C. i and iii only   D. i, ii and iii

26. For each of the statements given below, determine whether $x = 0$ is a solution.

    i. $x^2 - 3x = x - 3$

    ii. $2 + x \leq x$

    iii. $|x - 10| \geq 5$

A. i and ii only   B. ii and iii only   C. iii only   D. ii only

CLAST SKILL IV.C.1 *The student solves real-world problems involving the use of variables, aside from commonly-used geometric formulas.*

See page 98 of this book to learn about this CLAST skill.

27. An equation for centripetal force is $F = \dfrac{mv^2}{r}$.

A truck with a mass m = 1500 kilograms, traveling at a constant speed v = 10 meters per second, requires a force (F) of 3000 Newtons to stay on a circular track of radius r. What is the length of r?

A. 50 meters   B. 500 meters   C. 20 meters   D. .5 meters

28. Kinetic energy is computed from the formula $E = \dfrac{1}{2}mv^2$.

If a projectile of mass (m) 4 kg is traveling at a constant speed (v) has 20,000 Joules of kinetic energy (E), then how fast is the projectile traveling? (In this formula, the speed will be expressed in meters per second.)

A. 10,000 meters per second   B. 4,000 meters per second

C. 4 meters per second   D. 100 meters per second

29. The daily revenue (R, in thousands of dollars) from the manufacture of x tons of a commodity is determined by the formula $R = 24x - x^2$. Since the manufacturer can produce at most 12 tons of the commodity per day, x must be less than or equal to 12 in this formula.

How many tons of the commodity must be manufactured in order to produce $140 thousand in revenue?

A. 8.5 tons   B. 6.4 tons   C. 10 tons   D. 12 tons

CLAST Exercises for Section 7.1 of *Thinking Mathematically*.

CLAST SKILL I.C.6 *The student will find particular values of a function.*

See page 74 of this book to learn about this CLAST skill.

1. Find $f(3)$ given $f(x) = 2x^2 + x - 4$
A. 13   B. 17   C. 31   D. 35

2. Find $f(-1)$ given $f(x) = x^3 - x^2 + x$
A. -3 B. -2 C. 0   D. -1

3. Find $f\left(\dfrac{1}{2}\right)$ given $f(x) = 16x^2 + 1$

A. $\dfrac{3}{4}$ B. 4   C. 9   D. 5

4. Find $f\left(\dfrac{5}{3}\right)$ given $f(x) = 6x - 3$

A. 2   B. 7   C. 1   D. 6

5. Find $f(4)$ given $f(x) = -x^2 + 2x$
A. 24   B. 0   C. -8 D. -12

6. Find $f\left(-\dfrac{1}{3}\right)$ given $f(x) = 18x^2 - 12x - 3$

A. -1 B. 3   C. -9 D. -5

7. Find $f(-2)$ given $f(x) = 2x^3 + x - 4$
A. -22 B. 2   C. 10   D. 14

8. Find $f\left(\dfrac{2}{3}\right)$ given $f(x) = 4 - 3x - 9x^2$

A. 1   B. 0   C. -2 D. 2

9. Find $f(-1)$ given $f(x) = 4x^3 + 2x^2 - 3x + 2$
A. 11   B. -3 C. -7 D. 3

10. Find $f(-4)$ given $f(x) = 6 - x^2$
A. 22   B. -22 C. -10 D. 4

CLAST Exercises for Section 7.3 of *Thinking Mathematically.*

CLAST SKILL I.C.9 *The student will solve a system of two linear equations in two unknowns.*

See page 79 of this book to learn about this CLAST skill.

1. Choose the correct solution set for the system of linear equations.
$3x + 5y = 17$
$x + 3y = 11$

A. $\{(-1, 4)\}$

B. $\left\{\left(7, -\frac{4}{5}\right)\right\}$

C. $\left\{(x, y) \mid y = -\frac{1}{5}(x + 11)\right\}$

D. The empty set

2. Choose the correct solution set for the system of linear equations.
$6x + 3y = -3$
$2x - 4y = 9$

A. $\{(1, -3)\}$

B. $\{(0, -1)\}$

C. $\left\{\left(\frac{1}{2}, -2\right)\right\}$

D. The empty set

3. Choose the correct solution set for the system of linear equations.
$-x + 4y = 3$
$2x - 8y = -4$

A. $\{(1, 1)\}$

B. $\{(2, -1)\}$

C. $\left\{(x, y) \mid y = \frac{1}{4}(x + 5)\right\}$

D. The empty set

4. Choose the correct solution set for the system of linear equations.
$5x - 2y = 3$
$3x - 4y = -15$

A. $\left\{\left(\frac{1}{5}, -1\right)\right\}$

B. $\{(3, 6)\}$

C. $\{(-1, -4)\}$

D. The empty set

SECTION 7.3

5. Choose the correct solution set for the system of linear equations.

$x - y = 1$
$4x + 2y = 7$

A. $\left\{\left(\frac{3}{2}, \frac{1}{2}\right)\right\}$

B. $\{(3, 6)\}$

C. $\{(x,y) \mid y = x + 1\}$

D. The empty set

6. Choose the correct solution set for the system of linear equations.

$2x - 4y = 6$
$-x + 2y = -3$

A. $\{(1, -1)\}$

B. $\left\{(x,y) \mid y = \frac{1}{2}(x - 3)\right\}$

C. $\{(-1, 1)\}$

D. The empty set

7. Choose the correct solution set for the system of linear equations.

$-x + 3y = -1$
$2x + 6y = 6$

A. $\left\{\left(-\frac{1}{2}, -\frac{2}{3}\right)\right\}$

B. $\left\{\left(\frac{1}{2}, 0\right)\right\}$

C. $\left\{\left(2, \frac{1}{3}\right)\right\}$

D. The empty set

8. Choose the correct solution set for the system of linear equations.

$10x + 4y = 2$
$3x - 2y = -1$

A. $\left\{\left(\frac{1}{4}, -\frac{1}{8}\right)\right\}$

B. $\{(-1, 2)\}$

C. $\left\{\left(0, \frac{1}{2}\right)\right\}$

D. $\left\{(x,y) \mid y = \frac{1}{2}(5x + 1)\right\}$

226

9. Choose the correct solution set for the system of linear equations.
$5x + 3y = 9$
$4x + 2y = 4$

A. $\left\{\left(1,\dfrac{4}{3}\right)\right\}$

B. $\{(-3, 8)\}$

C. $\{(x.y) \mid y = -2x + 2\}$

D. The empty set

10. Choose the correct solution set for the system of linear equations.
$2x + 3y = 3$
$10x + 15y = 4$

A. $\left\{\left(1,\dfrac{1}{3}\right)\right\}$

B. $\{(0,1)\}$

C. $\left\{(x,y) \mid y = \dfrac{1}{3}(-2x - 1)\right\}$

D. The empty set

11. Choose the correct solution set for the system of linear equations.
$6x + 3y = 0$
$4x + 4y = 4$

A. $\{(0,1)\}$

B. $\{(x,y) \mid y = -2x\}$

C. $\{(-1,2)\}$

D. The empty set

12. Choose the correct solution set for the system of linear equations.
$-3x + 2y = 1$
$18x - 12y = -6$

A. $\{(-1,-2)\}$

B. $\left\{(x,y) \mid y = \dfrac{1}{2}(3x + 1)\right\}$

C. $\{(5,7)\}$

D. The empty set

CLAST Exercises for Section 7.4 of *Thinking Mathematically.*

CLAST SKILL II.C.4 *The student will identify regions of coordinate plane which correspond to specified conditions and vice versa.*

See page 90 of this book to learn about this CLAST skill.

1. Select the conditions corresponding to the shaded region of the plane.

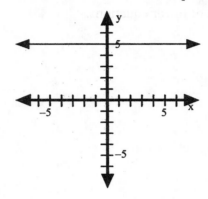

A. $x < 5$ B. $y < 5$

C. $x \leq 5$ D. $y \leq 5$

2. Select the conditions corresponding to the shaded region of the plane shown below.

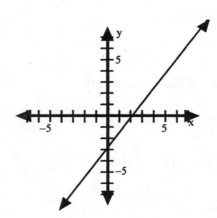

A. $3x - 2y \geq 6$ B. $3x - 2y \leq 6$

C. $2x + 3y \leq 6$ D. $2x + 3y \geq 6$

3. Select the figure whose shaded region corresponds to the conditions $x + 3y > -3$.

A.

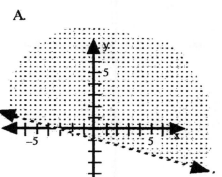

B.

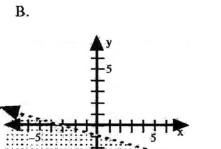

C.

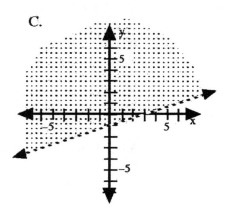

D.

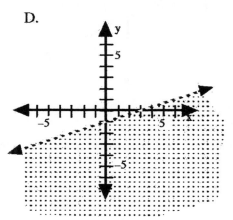

4. Select the graph whose shaded region corresponds to the conditions $x + 2y \leq 0$

A.

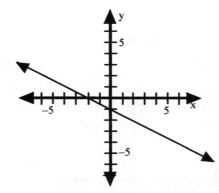

B.

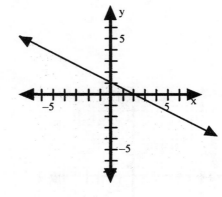

C.

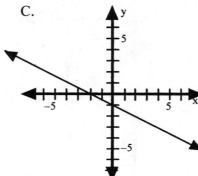

D.

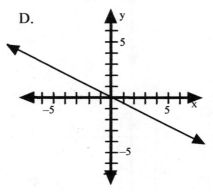

5. Select the conditions corresponding to the shaded region shown below.

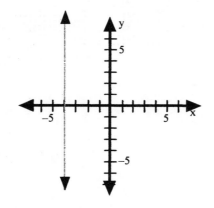

A. $y < -4$ B. $x \leq -4$

C. $x > -4$ D. $x < -4$

6. Select the graph whose shaded region corresponds to the conditions $x \leq 5$ and $y \geq 2$.

A.

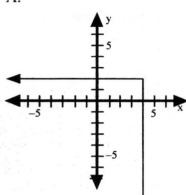

B.

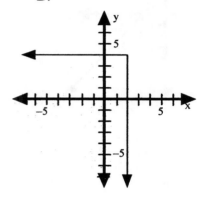

C.

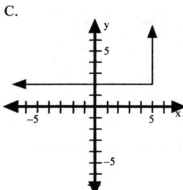

D.

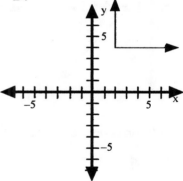

7. Select the graph whose shaded region corresponds to the conditions $x + y \geq 2$ and $y \leq 1$.

A.

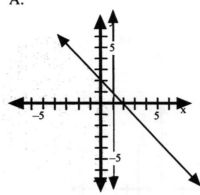

B.

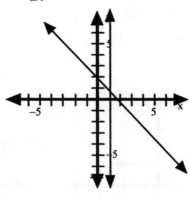

C.

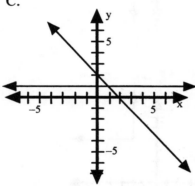

D.

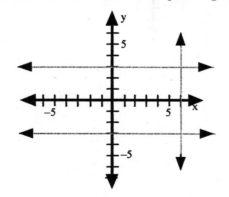

8. Select the conditions corresponding to the shaded region shown below.

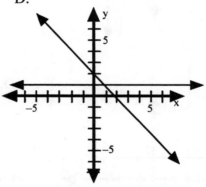

A. $x \leq 6$ and $y \leq 3$ and $y \geq -3$

B. $y \leq 6$ and $x \leq 3$ and $x \geq -3$

C. $y < 6$ and $x < 3$ and $x > -3$

D. $x < 6$ and $y < 3$ and $y > -3$

232

EXERCISES

9. Select the graph whose shaded region corresponds to the conditions
$2x - 5y \le 4$ or $x + y \le -2$.

A.

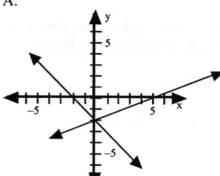

B.

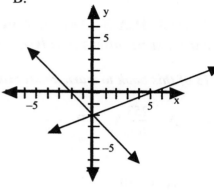

C.

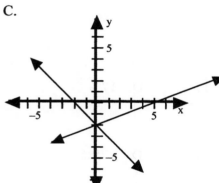

D.

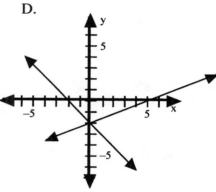

10. Select the conditions corresponding to the shaded region of the plane shown below.

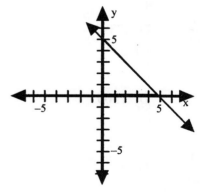

A. $x \le 5$ and $y \le 5$

B. $x \le 5$ or $y \le 5$

C. $x + y \le 5$ and $x \ge 0$ and $y \ge 0$

D. $x + y \le 5$ or $x \ge 0$ or $y \ge 0$

233

CLAST Exercises for Section 8.1 of *Thinking Mathematically.*

CLAST SKILL II.A.3 *The student will identify equivalent forms of positive rational numbers involving decimals, percents and fractions.*

See page 13 of this book to learn about this CLAST skill.

1. $0.253 =$ A. $\dfrac{253}{1000}\%$ B. $\dfrac{253}{100}$ C. 253% D. $\dfrac{253}{1000}$

2. $180\% =$ A. 0.180 B. 1.80 C. 180.0 D. 1800

3. $\dfrac{11}{20} =$ A. 55 B. 55% C. 5.5 D. $.55\%$

4. $\dfrac{85}{500} =$ A. 85% B. $.85$ C. $.17$ D. $.17\%$

5. $1\dfrac{3}{4} =$ A. $.75$ B. $.75\%$ C. 1.75 D. 1.75%

6. $60 =$ A. 6000% B. 60% C. 600% D. $.60$

7. $0.739 =$ A. 739% B. $\dfrac{739}{100}$ C. $\dfrac{739}{1000}$ D. $.739\%$

8. $0.15\% =$ A. 15 B. $\dfrac{15}{100}$ C. $.0015$ D. 1500

9. $310\% =$ A. $3\dfrac{1}{10}$ B. $.310$ C. $\dfrac{310}{1000}$ D. 31

10. $0.4682 =$ A. 4.682% B. $\dfrac{10000}{4682}$ C. $\dfrac{4682}{1000}$ D. 46.82%

11. $0.009 =$ A. $\dfrac{9}{1000}$ B. $.09\%$ C. 9% D. $\dfrac{9}{100}$

12. $10\dfrac{2}{5} =$ A. 10.4 B. 10.4% C. 10.04 D. 10.2

CLAST SKILL I.A.4 *The student will solve the sentence a% of b is c, where values for two of the variables are given.*

See page 7 of this book to learn about this CLAST skill.

13. What is 20% of 8.25?

A. 165 B. 1.65 C. 9.9 D. 5.25

14. 3.6 is 4% of what number?

A. 90 B. 14.4 C. 144 D. 111

15. 12 is what percent of 32?

A. .375% B. 37.5% C. 23.3% D. 233%

16. What is 2.4% of 80?

A. $3\dfrac{1}{3}$ B. $33\dfrac{1}{3}$ C. 1.92 D. 19.2

17. 15 is 20% of what number?

A. 3 B. 30 C. $1\dfrac{3}{4}$ D. 75

18. What is 0.4% of 200?

A. 5 B. .8 C. .5 D. 8

19. 60 is what percent of 30?

A. 50% B. 200% C. 20% D. 180%

SECTION 8.1

20. What is 2% of 20?

   A. 1000   B. 10   C. 4   D. .4

21. 12 is 15% of what number?

   A. 180   B. 1.8   C. 80   D. 18

22. 48 is 120% of what number?

   A. 57.6   B. 40   C. 25   D. 250

23. Find 55% of 280.

   A. 154   B. 5.09   C. 509   D. 154,000

24. 22.5 is what percent of 75?

   A. .3%   B. 30%   C. 168.75% D. 16.875%

CLAST SKILL I.A.3 *The student will demonstrate the ability to calculate percent increase and percent decrease.*

See page 5 of this book to learn about this CLAST skill.

25. If 8 is increased to 10, what is the percent increase?

   A. 20%   B. 25%   C. 80%   D. 37%

26. If 20 is increased by 30% of itself, what is the result?

   A. 23   B. 50   C. 6   D. 26

27. If 150 is decreased to 75, find the percent decrease.

   A. 50%   B. 67%   C. 75%   D. 100%

28. If 22 is increased by 100% of itself, what is the result?

A. 44   B. 122   C. 24.2   D. 144

29. If 120 is decreased to 102, what is the percent decrease?

A. 102%   B. 18%   C. 15%   D. 20%

30. If 250 is decreased by 20% of itself, what is the result?

A. 230   B. 237.5   C. 50   D. 200

31. If 300 is increased to 354, what is the percent increase?

A. 84.4%   B. 18%   C. 54%   D. 15%

32. If 160 is decreased to 48, find the percent decrease.

A. 112%   B. 7%   C. 70%   D. 23%

33. If 450 is increased to 459, what is the percent increase?

A. 98%   B. 5%   C. 9%   D. 2%

34. If 860 is decreased by 40% of itself, what is the result?

A. 516   B. 344   C. 21.5   D. 215

CLAST SKILL IV.A.2 *The student solves real-world problems which do not require the use of variables but do require the use of percent.*

See page 25 of this book to learn about this CLAST skill.

35. Adrian works for a note-taking service. In order to determine the demand for published class notes in a biology course, he has conducted a survey and received responses from 15% of the students enrolled in the course. The actual number of students he has contacted is 30. How many students are enrolled in the course?

A. 250   B. 480   C. 450   D. 200

SECTION 8.1

36. If the cost of a portable TV is reduced from $80 to $60, what is the percent decrease in cost?

A. 30%   B. 25%   C. 20%   D. 35%

37. Last year Anne's GPA was 3.40. This year her GPA has increased by 10%. What is her GPA this year?

A. 3.30   B. 3.34   C. 13.40   D. 3.74

38. On the first day of class, there were 40 students in the class Rocket Science 101. By the end of the second day of class, there were only 10 students left. Find the percent decrease in the number of students.

A. 30%   B. 25%   C. 75%   D. 33.3%

39. Last year, the property taxes on Juan's home were $520. This year, the property taxes will increase by 3%. Find the amount of increase.

A. $15.60   B. $156   C. $535.60   D. $676

40. Before he started using Monstro's Miracle Engine Additive, Jed's truck used two quarts of oil per week. The dealer claimed that the engine additive would reduce oil consumption by 80%. If the claim is true, what will be Jed's weekly oil consumption?

A. 1.6 quarts   B. 1.2 quarts   C. .8 quarts   D. .4 quarts

41. Researchers expect that 16% of adults will have scores greater than 115 on a certain IQ measurement test. If 400 people take the test, how many would be expected to have IQs greater than 115?

A. 46   B. 64   C. 25   D. 3.48

CLAST SKILL I.D.1 *The student will identify information contained in bar, line and circle graphs.*

See page 103 of this book to learn about this CLAST skill.

42. A school district purchased 250 personal computers. This make of computer is available in a number of different colors, or "flavors." The pie chart below shows the distribution according to "flavor" of the computers in this purchase. How many were not "bluebird?"

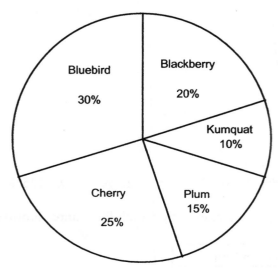

A. 30   B. 70   C. 175   D. 75

43. The bar graph below shows the distribution of grades in a European History course.

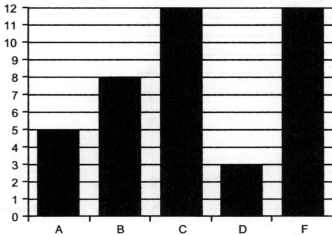

What percent of students had grades less than "C?"
A. 15%   B. 37.5%   C. 30%   D. 53%

SECTION 8.1

44. A number of people were asked what animal they would prefer for a pet. Their responses are summarized in the pie chart below; each person was allowed only one response.

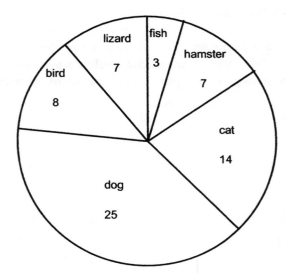

What percent responded "bird?"   A. 8% B. 51.2%   C. 12.5%   D. .125%

45. The line graph below shows the average temperature by month for Tampa, Florida.

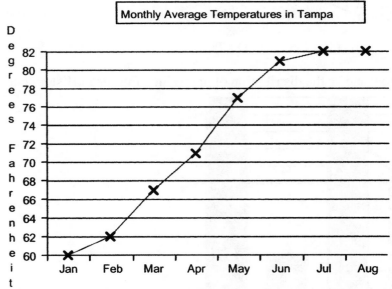

The average monthly temperature in February is roughly what percent greater than the average monthly temperature in January?

A. 2%   B. 62%   C. 3.33%   D. 33.3%

CLAST Exercises for Section 9.1 of *Thinking Mathematically.*

CLAST SKILL I.B.1 *The student will round measurements to the nearest given unit of the measuring device.*

See page 29 of this book to learn about this CLAST skill.

1. Round the measure 24.49 centimeters to the nearest centimeter.

A. 24 cm

B. 24.5 cm

C. 25 cm

D. 245 cm

2. Round the measure 872.58 millimeters to the nearest tenth of a millimeter.

A. 872 mm

B. 872.5 mm

C. 872.6 mm

D. 873 mm

3. Round 18.997 feet to the nearest hundredth of a foot.

A. 19.00 ft

B. 18.90 ft

C. 18.99 ft

D. 19.09 ft

4. Round the measure 19.8 inches to the nearest foot.

A. 20 ft

B. 2 ft

C. 19 ft

D. 1 ft

5. Round the measure 158.4 inches to the nearest yard.

A. 160 yd

B. 158 yd

C. 159 yd

D. 4 yd

SECTION 9.1

6. Round the measurement of the nail to the nearest half-inch.

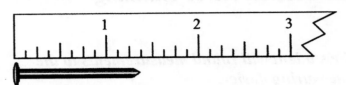

A. 1 in

B. $1\frac{1}{2}$ in

C. 2 in

D. $\frac{1}{2}$ in

7. Round the diameter of the washer to the nearest 1/4-inch.

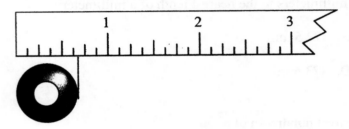

A. 1/4 inch

B. 1/2 inch

C. 3/4 inch

D. 1 inch

8. Round the measurement 183.72 inches to the nearest ten inches.

A. 184 inches

B. 183.7 inches

C. 18 inches

D. 180 inches

CLAST Exercises for Section 9.2 of *Thinking Mathematically.*

CLAST SKILL I.B.1 *The student will round measurements to the nearest given unit of the measuring device.*

See page 29 of this book to learn about this CLAST skill.

1. Round the measure 806.05 cubic centimeters to the nearest cubic centimeter.

A. 81 cu. cm   B. 806 cu. cm   C. 807 cu. cm   D. 80 cu. cm

2. Round the measure 16.59 square inches to the nearest tenth of a square inch.

A. 16.0 sq. in   B. 17.0 sq. in   C. 16.6 sq. in   D. 16.5 sq. in

3. Round the measure 645.02 acres to the nearest 10 acres.

A. 64.5 acres   B. 6.45 acres   C. 640 acres   D. 650 acres

CLAST SKILL II.B.4   *The student will identify appropriate units of measure for geometric objects.*

See page 47 of this book to learn about this CLAST skill.

4. Identify the units of measure that would be appropriate for measuring the diameter of a basketball.

A. centimeters   B. square inches

C. cubic millimeters   D. cubic inches

5. Identify the units of measure that would be appropriate for measuring the amount of propane in a cylindrical propane tank.

A. square yards   B. meters

C. centimeters   D. cubic feet

SECTION 9.2

6. Identify the units of measure that would be appropriate for measuring the distance around a rectangular garden plot.

A. square feet

B. meters

C. cubic centimeters

D. cubic yards

7. Identify the units of measure that would be appropriate for measuring the amount of fabric needed to cover a rectangular wall.

A. square yards   B. cubic feet   C. millimeters   D. meters

8. Identify the units of measure that would be appropriate for measuring the amount of flat material needed to construct a rectangular box.

A. meters

B. cubic meters

C. cubic feet

D. square inches

9. Identify the units of measure that would be appropriate for measuring the thickness of a nail.

A. square millimeters

B. cubic centimeters

C. millimeters

D. square centimeters

10. Identify the units of measure that would be appropriate for measuring the amount of fluid contained in a spherical vessel.

A. feet

B. yards

C. square meters

D. cubic inches

11. Identify the units of measure that would be appropriate for measuring the length of a string stretching from one corner of a rectangular floor to the opposite corner.

A. cubic centimeters

B. square yards

C. meters

D. cubic feet

CLAST Exercises for Section 10.1 of *Thinking Mathematically.*

CLAST SKILL II.B.2 *The student will classify simple plane figures by recognizing their properties.*

See page 40 of this book to learn about this CLAST skill.

1. Referring to the diagram below, which of the following pairs of angles are complementary?

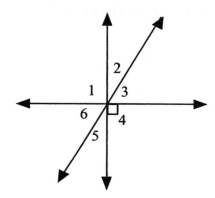

A. 2 and 3   B. 1 and 2   C. 6 and 3   D. 4 and 5

2. Referring to the diagram below, which of the following pairs of angles are vertical?

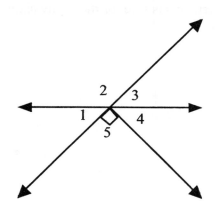

A. 4 and 3   B. 2 and 4   C. 1 and 3   D. 2 and 3

3. Referring to the diagram below, which angle is acute?

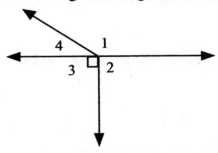

A. 1   B. 2   C. 3   D. 4

4. Referring to the figure below, select the angle that is supplementary to angle 2.

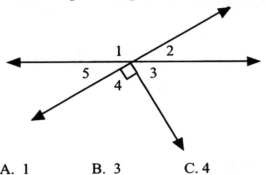

A. 1   B. 3   C. 4   D. 5

CLAST SKILL II.B.1   The student will identify relationships between angle measures.

See page 37 of this book to learn about this CLAST skill.

5. Given that $\overline{AB} \parallel \overline{CD}$, which of the following statements is true for the figure shown? (The measure of angle CBA is represented by "x.")

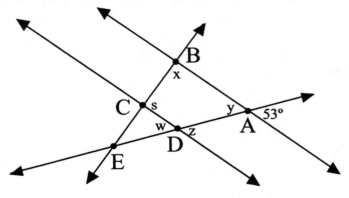

A. y = 37°   B. x = y   C. w = 53°   D. s = w

6. Which of the following statements is true for the figure shown? (The measure of angle BAC is represented by "x.")

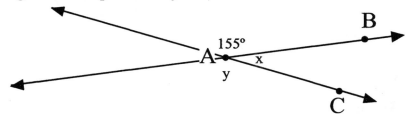

A. $x = y$ B. $y = 155°$ C. $x = 155°$ D. $x = 45°$

7. Given that $\overline{AC} \parallel \overline{BD}$, which of the following statements is true for the figure shown? (The measure of angle CAB is represented by "x.")

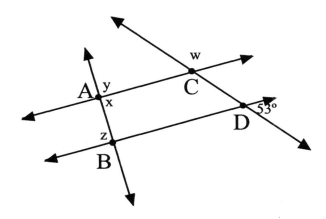

A. $w = 53°$ B. $x = 37°$ C. $y = z$ D. $z = x$

8. Given that $\overline{AB} \parallel \overline{CD}$, which of the following statements is true for the figure shown? (The measure of angle BCE is represented by "x.")

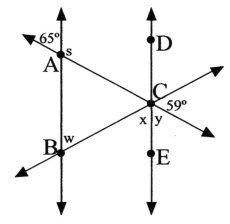

A. $w = y$ B. $x = s$ C. $w = 56°$ D. $x = 59°$

9. Given that $\overline{AC} \parallel \overline{BD}$, which of the following statements is true for the figure shown? (The measure of angle ABD is represented by "x.")

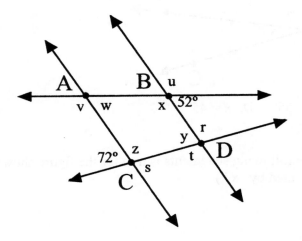

A. r = 38°

B. r = 108°

C. t = 52°

D. w = 72°

CLAST Exercises for Section 10.2 of *Thinking Mathematically.*

CLAST SKILL II.B.2 *The student will classify simple plane figures by recognizing their properties.*

See page 40 of this book to learn about this CLAST skill.

1. What type of triangle is △ABC?

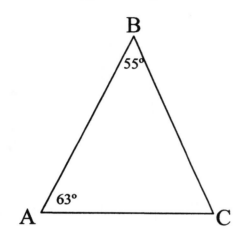

A. equilateral   B. obtuse

C. isosceles   D. scalene

2. What type of triangle is △ABC?

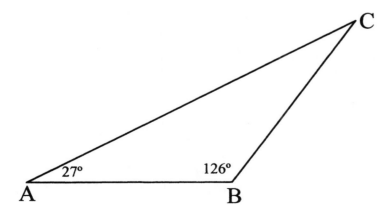

A. equilateral   B. acute   C. isosceles   D. scalene

3. Given that $\overline{AD} \parallel \overline{BC}$ in the figure below, what type of triangle is $\triangle ABC$?

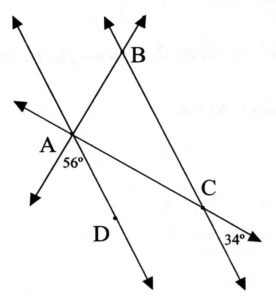

A. obtuse

B. right

C. isosceles

D. equilateral

CLAST SKILL II.B.1   The student will identify relationships between angle measures.

See page 37 of this book to learn about this CLAST skill.

4. Given that $\overline{AB} \parallel \overline{DC}$, which of the following statements is true for the figure shown? (The measure of angle EAB is represented by "x.")

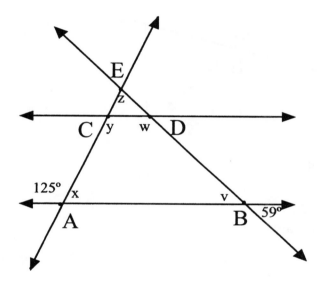

A. $y = w$

B. $\overline{EA} \cong \overline{EB}$

C. $z = 66°$

D. $w = 120°$

5. Given that $\overline{AD} \parallel \overline{BC}$, which of the following statements is true for the figure shown? (The measure of angle ABC is represented by "x.")

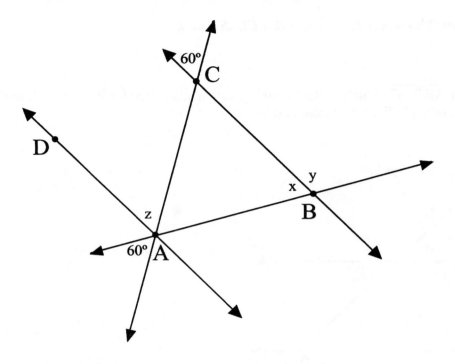

A. $\angle ABC$ and $\angle BCA$ are supplementary

B. $\triangle ABC$ is equilateral

C. $y = z$

D. $x = 120°$

6. Which of the following statements is true for the figure shown? (The measure of angle CAB is represented by "x.")

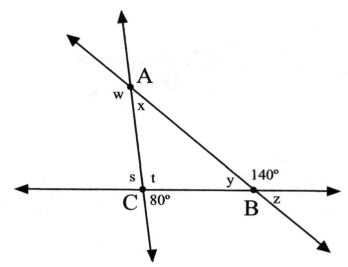

A. w = 80°

B. y = t

C. x = y

D. w = t

7. Select the statement that is true for the figure shown (The measure of angle BCD is represented by "x.)

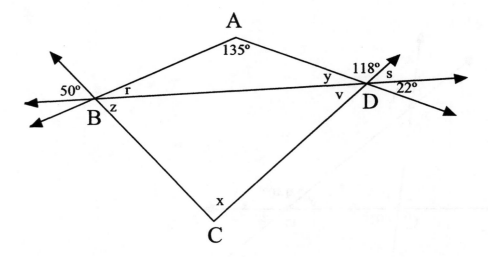

A. $r = 50°$

B. $y = v$

C. $\overline{AB} \parallel \overline{DC}$

D. $\overline{BC} \perp \overline{DC}$

8. Given that $\overline{AB} \parallel \overline{CD}$ and $\overline{AC} \parallel \overline{BD}$, which of the following statements is true for the figure shown? (The measure of angle ACD is represented by "x.")

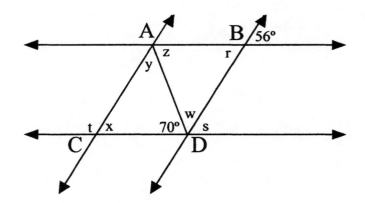

A. $w = s$ B. $\overline{AD} \perp \overline{AC}$ C. $t = 124°$ D. $s = 70°$

CLAST SKILL II.B.3 *The student will recognize similar triangles and their properties.*

See page 42 of this book to learn about this CLAST skill.

9. Which of the statements A – D is true for the pictured triangles?

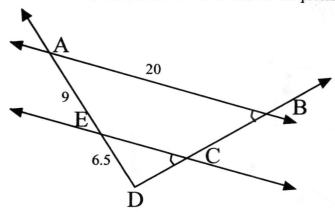

A. $\dfrac{AB}{BD} = \dfrac{DC}{EC}$　　　B. $\dfrac{9}{20} = \dfrac{6.5}{EC}$

C. $\dfrac{DC}{DB} = \dfrac{DE}{DA}$　　　D. $\dfrac{20}{DB} = \dfrac{DC}{AE}$

10. Which of the statements A – D is true for the pictured triangles?

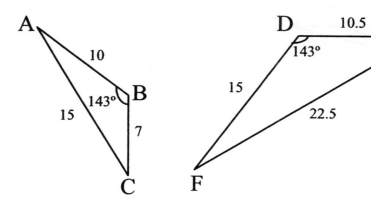

A. $\dfrac{AB}{BC} = \dfrac{ED}{FD}$　　　B. $\angle CAB \cong \angle FED$

C. $\angle BCA \cong \angle DEF$　　　D. $\dfrac{10}{15} = \dfrac{DE}{DF}$

SECTION 10.2

11. Select the figure below in which all triangles are similar.

A.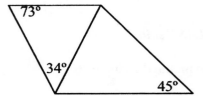

B.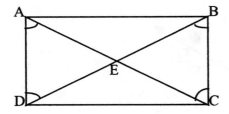

C.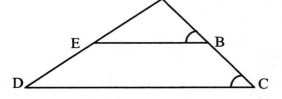

D.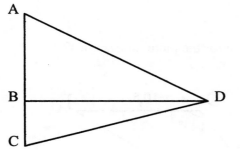

12. Which of the statements A – D is true for the pictured triangles?

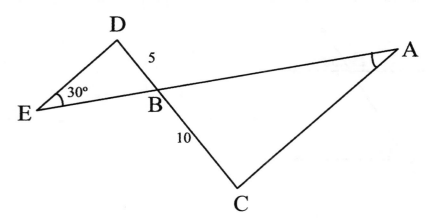

A. $\dfrac{AB}{AC} = \dfrac{ED}{EB}$ B. $\dfrac{AB}{EB} = \dfrac{AC}{BC}$

C. $\dfrac{5}{10} = \dfrac{CB}{AB}$ D. $\dfrac{BE}{5} = \dfrac{AB}{10}$

13. Given that $\overline{AB} \parallel \overline{CD}$, which of the statements A – D is true for the pictured triangles?

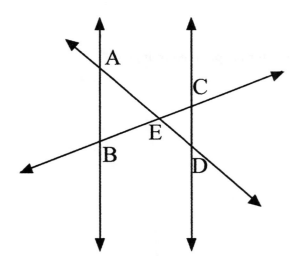

A. $\angle BAE \cong \angle DCE$ B. $\dfrac{CD}{BA} = \dfrac{CE}{BE}$

C. $\dfrac{CE}{DE} = \dfrac{AB}{BE}$ D. $\dfrac{AE}{BE} = \dfrac{DE}{DC}$

SECTION 10.2

14. Given that $\overline{AB} \parallel \overline{CD}$, which of the statements A – D is true for the pictured triangles?

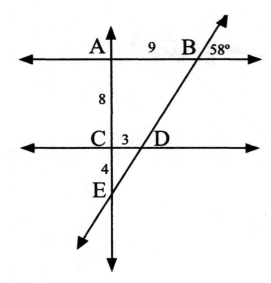

A. $\dfrac{AE}{CE} = \dfrac{AB}{CD}$　　　B. $\dfrac{EB}{ED} = \dfrac{DC}{BA}$

C. $\dfrac{3}{4} = \dfrac{9}{8}$　　　D. $\dfrac{AE}{AB} = \dfrac{3}{4}$

15. Which of the statements A – D is true for the pictured triangles?

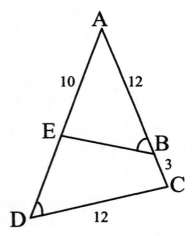

A. $\dfrac{AC}{AD} = \dfrac{AB}{AE}$　　　B. $AD = 12.5$

C. $EB = 8$　　　D. $\angle ACD \cong \angle ABE$

258

16. Study figures A through D below. Select the figure in which all of the triangles are similar.

A.

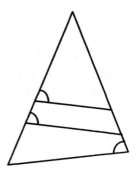

B.

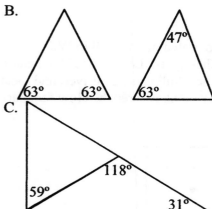

C.

D.

SECTION 10.2

CLAST SKILL IV.B.2 *The student solves real-world problems involving the Pythagorean property.*

See page 61 of this book to learn about this CLAST skill.

17. Two boats leave port at the same time, one of them traveling northward at a constant speed and the other traveling westward at a constant speed. After 3 hours the first boat is 24 miles north of the port and 30 miles from the other sailboat. How fast is the second sailboat traveling?

A. 18 miles per hour   B. 6 miles per hour

C. 10 miles per hour   D. 13 miles per hour

18. The drawing below shows a pegboard that is mounted on the wall in Wilma's workshop. The holes in the pegboard are three inches apart. Location "X" indicates the hole in the pegboard from which Wilma will hang her Phillips screwdriver, and location "Y" indicates the hole from which she will hang her monkey wrench. How far apart are the two holes?

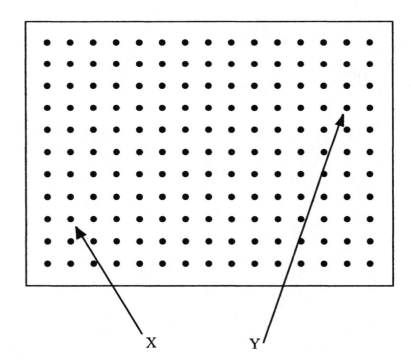

A. 13 inches   B. 17 inches

C. 39 inches   D. 51 inches

19. On three days per week an auto-parts delivery person makes the trip from Tinyberg to Littleton on County Road A, then takes the Littleton Parkway to Junction City and returns to Tinyberg on County Road YY (see map below). Find the total distance traveled on one of these trips.

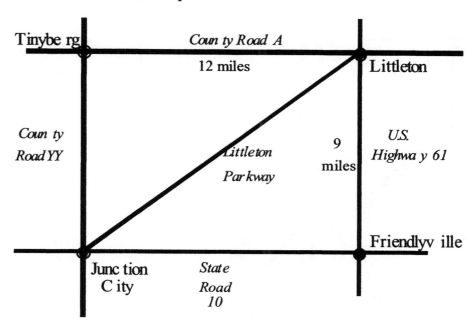

A. 36 miles   B. 15 miles   C. 57 miles   D. 42 miles

20. Alton's flagpole has a height of 16 yards. It will be supported by three cables, each of which is attached to the flag pole at a point 4 yards below the top of the flag pole and attached to the ground at a place that is 9 yards from the base of the flag pole. Find the total number of feet of cable that will be required.

A. 45 feet   B. 15 feet

C. 60 feet   D. 135 feet

21. A rocket ascends vertically after being launched from a location that is midway between two ground-based tracking stations. When the rocket reaches an altitude of 4 kilometers it is 5 kilometers from each of the tracking stations. How far apart are the two tracking stations? Assume that this is a locale where the terrain is flat.

A. 3 km   B. 4 km

C. 5 km   D. 6 km

SECTION 10.2

22. A stained glass worker has two rectangular sheets of glass that each measure 6 inches wide and 8 inches long. The sheets will each be cut along the diagonal, and the four resulting triangular pieces will be recombined to form a "kite" shape (see figure below). The kite shape will be bordered by a perimeter formed from a metal strip. What will be the length of the metal strip?

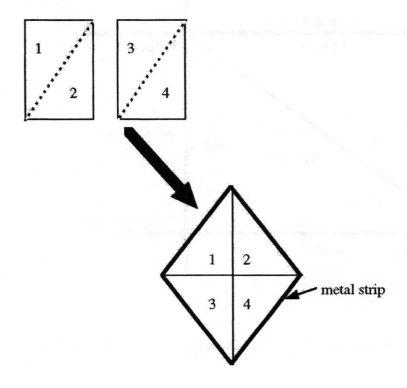

A. 40 feet

B. 10 feet

C. $3\frac{1}{3}$ feet

D. $2\frac{1}{3}$ feet

23. Harry has a rectangular garden bed measuring 5 feet by 12 feet. A water faucet is located at one corner of the garden bed. A hose will be connected to the water faucet. Harry wants the hose to be long enough to reach the opposite corner of the garden bed when stretched straight. Find the required length of hose.

A. 13 feet

B. 15 feet

C. 17 feet

D. 21 feet

24. In order to measure the distance from point A on one side of a swamp to point B on the other side of the swamp, a surveyor marks off the right triangle shown in the figure below. What is the distance from point A to point B?

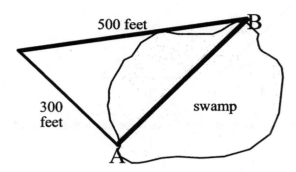

A. 300 feet   B. 500 feet   C. 583 feet   D. 400 feet

CLAST SKILL III.B.1 *The student infers formulas for measuring geometric figures.*

See page 49 of this book to learn about this CLAST skill.

25. Study the figures below, each of which provides information about the side length and height of a triangle.

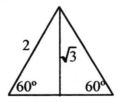 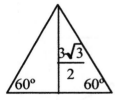 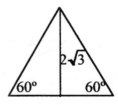 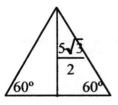

Find the height of a similar triangle whose side length is 6.

A. $\dfrac{\sqrt{3}}{6}$

B. $6\sqrt{3}$

C. $\dfrac{\sqrt{3}}{2}$

D. $3\sqrt{3}$

SECTION 10.2

26. Study the figures below, each of which provides information about the measure of the base and the measure of the hypotenuse of a right triangle.

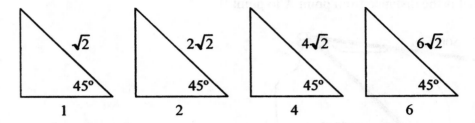

Find the measure of the hypotenuse of a similar triangle whose base has a measure of 10.

A. $3\sqrt{10}$ B. $2\sqrt{10}$

C. $10\sqrt{2}$ D. $10 + \sqrt{2}$

27. Study the figures below, each of which provides information about the measure of the height and the perimeter (P) of a right triangle.

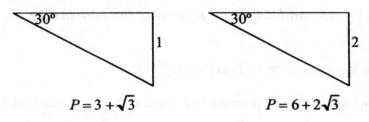

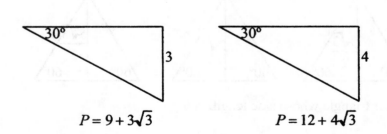

Find the perimeter of a similar triangle when the height is 8.

A. $16 + \sqrt{8}$ B. $24 + 3\sqrt{8}$

C. $8 + 24\sqrt{3}$ D. $24 + 8\sqrt{3}$

28. Study the figures below, each of which provides information about the measures of the three sides of a right triangle.

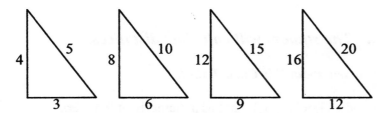

Find the measure of the hypotenuse of a similar triangle whose shorter leg measures 18 units.

A. 26   B. 30   C. 24   D. 25

CLAST Exercises for Sections 10.3 and 10.4 of *Thinking Mathematically.*

CLAST SKILL I.B.2a   *The student will calculate distances.*

See page 31 of this book to learn about this CLAST skill.

1. What is the distance around a circular pool that has a diameter of 9 meters?

A. 9π square meters   B. 9π meters

C. 81π square meters   D. 81π meters

2. What is the distance around the polygon shown below?

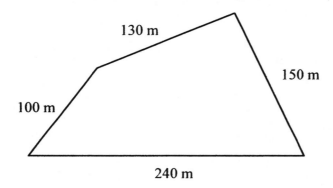

A. 62 km   B. 620 km   C. .62 km   D. 620,000 km

3. Find the distance around the regular octagon shown below.

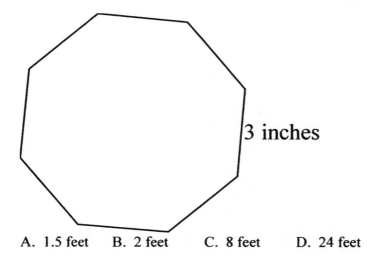

3 inches

A. 1.5 feet   B. 2 feet   C. 8 feet   D. 24 feet

4. What is the distance around a circular region with a radius of 10 yards?

A. 10π yards B. 5π yards C. 20π yards D. 25π yards

5. Find the distance around the right triangle shown below.

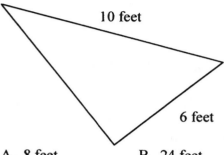

10 feet

6 feet

A. 8 feet B. 24 feet C. 4 feet D. 20 feet

6. What is the diameter of a circular flower garden if the distance around the flower garden is 36π yards?

A. 108 feet B. 18 feet C. 36 feet D. 72 feet

7. What is the distance around the regular pentagon shown below?

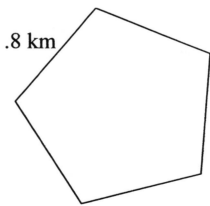

.8 km

A. 800 m

B. 40 m

C. 4 m

D. 4000 m

SECTION 10.4

CLAST SKILL I.B.2b   *The student will calculate areas.*

See page 33 of this book to learn about this CLAST skill.

8. Find the area of a circular region whose diameter is 12 inches.

A. 12π square inches

B. 12π inches

C. 36π square inches

D. 144π inches

9. What is the area of a square region whose sides measure 18 inches?

A. 1.5 square feet

B. 324 square feet

C. 27 square feet

D. 2.25 square feet

10. Find the area of the right triangle shown below.

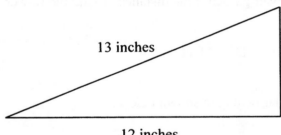

13 inches

12 inches

A. 30 square inches

B. 78 square inches

C. 5 square inches

D. 780 square inches

11. Find the area of the parallelogram shown below.

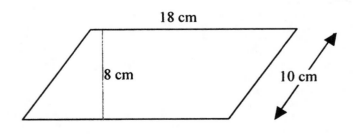

18 cm

8 cm

10 cm

A. 180 sq. cm

B. 144 sq. cm

C. 80 sq. cm

D. 1,440 sq. cm

12. Find the area of a rectangular region that is 12 meters wide and 8 meters long.

A. 96 meters   B. 40 meters   C. 40 sq. meters   D. 96 sq. meters

13. Find the area of a circular pond whose radius is 4 feet.

A. 16π sq. feet   B. 8π sq. feet   C. 4π sq. feet   D. 32π sq. feet

14. Find the area of the triangle shown below.

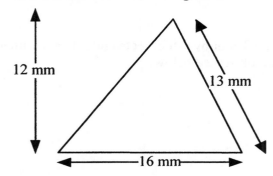

A. 192 sq. mm   B. 104 sq. mm   C. 96 sq. mm   D. 1248 sq. mm

15. Find the area of a circular region with a diameter of 24 inches.

A. 48π sq. feet   B. 2π sq. feet   C. 4π sq. feet   D. π sq. feet

16. Find the area of the parallelogram shown below.

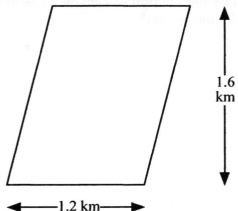

A. 1.92 sq. m   B. 1,920,000 sq. m   C. 1,920 sq. m   D. 19,200 sq. m

CLAST SKILL IV.B.1 *The student solves real world problems involving perimeters, areas and volumes of geometric figures.*

See page 57 of this book to learn about this CLAST skill.

17. A youth league baseball diamond is shaped like a square, with the bases and home plate located at the corners, and side lengths of 60 feet. How long will it take a player to run around the bases, starting and ending at home plate, if the player runs at an average speed of 4 yards per second?

A. 15 seconds   B. 20 seconds   C. 60 seconds   D. 180 seconds

18. An artist is going to completely cover a 3-foot by 4-foot rectangular board with a mosaic made of colored tiles. The tiles are illustrated below.

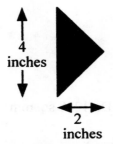

4 inches

2 inches

How many tiles are required to cover the board?

A. 96   B. 216   C. 432   D. 128

19. The diagram below shows the site plan for a proposed county park. The county commission is going to hire an architectural consulting firm to devise a landscaping program. For large-scale consulting projects like this, the firm will charge a rate of $600 per square mile. How much will they charge for this project?

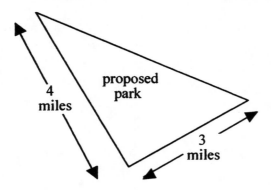

4 miles

proposed park

3 miles

A. $7,200   B. $3,600   C. $3,000   D. $36,000

20. The diagram below shows the site plan for a home. Mel's landscaping service will charge 2¢ per square yard for lawn maintenance. How much will this cost, assuming that the areas of the house, garage and drive (which measures 12 feet by 42 feet) will not be included?

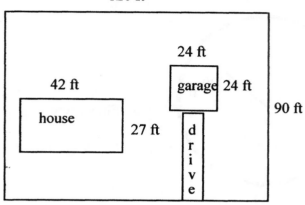

120 ft

24 ft

42 ft

garage 24 ft

90 ft

house

27 ft

d
r
i
v
e

A. $19.08   B. $57.24   C. $171.72   D. $44.92

21. A rectangular parcel of land measuring 300 yards by 240 yards will be enclosed by a fence costing 25¢ per foot. Find the total cost of the fence.

A. $162,000   B. $18,000   C. $810   D. $270

22. A homeowner is going to cover a rectangular 12-foot by 8-foot kitchen floor with square ceramic tiles measuring 6 inches by 6 inches. How many tiles are required?

A. 192   B. 384   C. 132   D. 132

23. Suppose that a circular pizza with a diameter of 6 inches cost $3.25. Assuming that the cost depends upon the size (area) of the pizza, find the cost of a similar pizza whose diameter is one foot.

A. $6.50

B. $19.50

C. $9.75

D. $13.00

24. The figure below shows a running track shaped like a figure-8. The runner starts at point A, circles around through points B and C, then through points D, E and F, back through point C, and then returns through point G to the starting point. If the runner completes the course 10 times, how many miles will he/she have run?

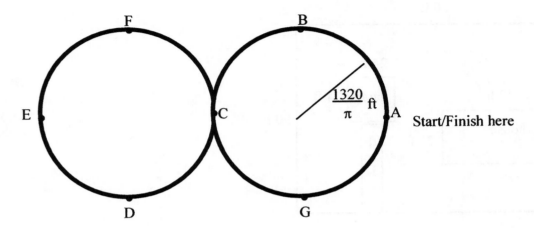

E ● Start/Finish here

A. 2.5

B. 5

C. 10

D. 20

25. A rectangular garden bed measuring 48 feet wide and 60 feet long will be bordered by a strip of cypress bark mulch that is one yard wide. How many square feet will be covered by the mulch?

A. 684 sq. ft

B. 218 sq. ft

C. 2368 sq. ft

D. 690 sq. ft

CLAST SKILL II.B.2 *The student will classify simple plane figures by recognizing their properties.*

See page 40 of this book to learn about this CLAST skill.

26. Identify the figure shown below.

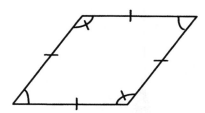

A. square

B. regular pentagon

C. rhombus

D. trapezoid

27. Select the figure that can simultaneously possess <u>all</u> of the following traits:

i. one pair of parallel sides;

ii. is a quadrilateral;

iii. opposite angles are not equal

A. parallelogram

B. rectangle

C. regular hexagon

D. trapezoid

28. Select the figure that must possess <u>all</u> of the following traits.

i. all sides have equal length;

ii. diagonals have equal length;

iii. opposite angles are equal;

A. rhombus

B. rectangle

C. square

D. trapezoid

CLAST SKILL III.B.1 *The student infers formulas for measuring geometric objects.*

See page 49 of this book to learn about this CLAST skill.

29. Study the information in the figures below.

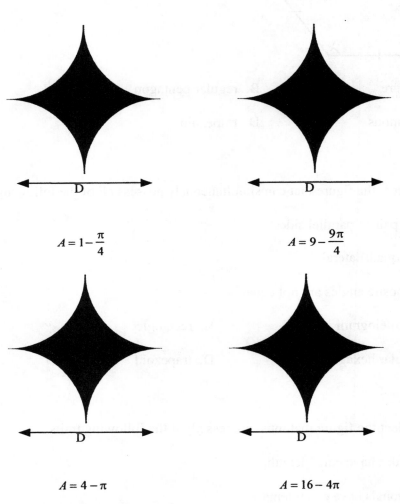

$$A = 1 - \frac{\pi}{4}$$

$$A = 9 - \frac{9\pi}{4}$$

$$A = 4 - \pi$$

$$A = 16 - 4\pi$$

Find A, the area of a similar figure when $D = 6$

A. $6 - 6\pi$ B. $6 - \frac{3\pi}{2}$ C. $36 - 9\pi$ D. $36 - \frac{9\pi}{4}$

30. Study the information in the figures below, each of which consists of a square from which one or more circles have been removed. In each figure, N represents the number of circles that have been removed, r represents the radius of each circle, and A represents the area of the shaded part of the figure.

N = 1

r = 1

A = 4 − š

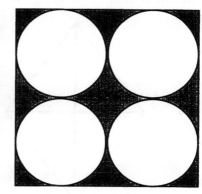

N = 4

r = 1/2

A = 4 − š

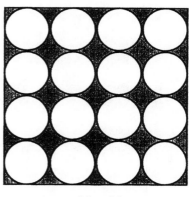

N = 16

r = 1/4

A = 4 − š

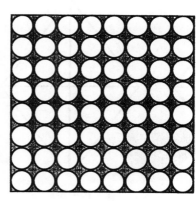

N = 64

r = 1/8

A = 4 − š

Find A, the area of the shaded part of a similar figure, if 128 circles have been removed.

A. 1/16

B. 128

C. 128 − 128π

D. 4 − π

31. Referring to the information in problem #30 above, find r if N = 256.

A. 16   B. 1/16   C. 1/32   D. 4 − 4π

32. Study the information in the following figures.

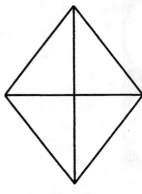

P = 20

A = 24

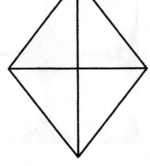

P = 40

A = 96

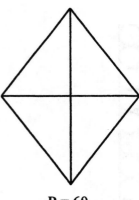

P = 60

A = 216

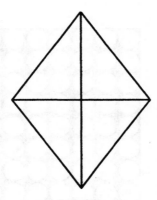

P = 80

A = 384

Find the area (A) of a similar figure if the perimeter (P) is 120.

A. 288

B. 432

C. 864

D. 1440

33. Study the information in the figures below. In each figure, A represents the area of the shaded part of the figure.

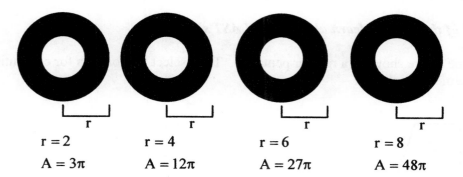

r = 2   r = 4   r = 6   r = 8
A = 3π   A = 12π   A = 27π   A = 48π

Find the area (A) of a similar figure if r = 14.

A. 196π

B. 109π

C. 147π

D. 21π

34. Study the given information. In each figure, S represents the sum of the measures of the interior angles.

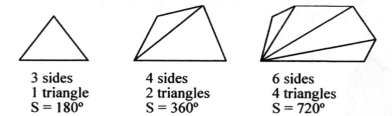

3 sides   4 sides   6 sides
1 triangle   2 triangles   4 triangles
S = 180°   S = 360°   S = 720°

Find S, the sum of the measures of the interior angles of a nine-sided convex polygon.

A. 1180°

B. 1260°

C. 1440°

D. 1620°

CLAST SKILL III.B.2 *The student identifies applicable formulas for computing measures of geometric figures.*

See page 54 of this book to learn about this CLAST skill.

35. Study the figure showing a regular pentagon. Then select the formula for computing the total area of the pentagon.

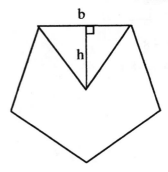

A. Area = (5/2)bh

B. Area = 5bh

C. Area = 5b + 2h

D. Area = 5(h + b)

36. Study the figure showing two semicircles attached to a rectangle, and then select the formula for calculating the area of the figure.

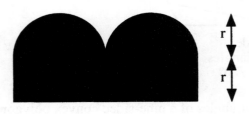

A. Area $= 2\pi r^2 + 5r^2$

B. Area $= \pi r^2 + 4r^2$

C. Area $= \pi r^2 + 5r^2$

D. Area $= 2\pi r^2 + 4r^2$

37. Study the figure below, which shows two squares joined to a rhombus, and then select the formula for calculating the area of the figure.

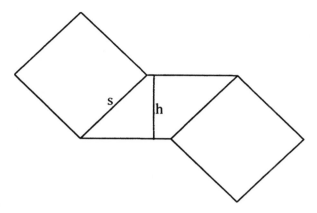

A. Area = 8s

B. Area = 10s + h

C. Area = 3s²h

D. Area = 2s² + sh

38. Study the figure below, which shows a semicircle attached to a square, and then select the formula for computing the perimeter of the figure.

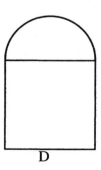

A. $\text{perimeter} = 3D + \frac{1}{2}\pi D$ B. $\text{perimeter} = 3D + \pi D$

C. $\text{perimeter} = D^2 + \frac{1}{8}\pi D^2$ D. $\text{perimeter} = D^2 + \frac{1}{2}\pi D^2$

39. Study the figure below, showing a rectangle to which an isosceles right triangle has been joined, and then select the formula for calculating the area of the figure.

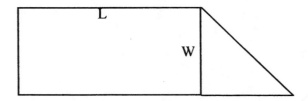

A. Area $= LW + W^2$

B. Area $= LW^2$

C. Area $= LW + (1/2)W^2$

D. Area $= 2L + 2W + W\sqrt{2}$

40. Study the figure below, showing two isosceles right triangles, and then select the formula for calculating the perimeter of the figure.

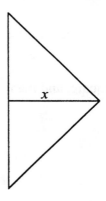

A. perimeter $= 6x$

B. perimeter $= 2x + 2x\sqrt{2}$

C. perimeter $= x^2$

D. perimeter $= 4x$

CLAST Exercises for Section 10.5 of *Thinking Mathematically.*

CLAST SKILL I.B.2b *The student will calculate areas.*

CLAST SKILL I.B.2c *The student will calculate volumes.*

See page 35 of this book to learn about this CLAST skill.

1. Find the surface area of a rectangular solid measuring 1 foot wide, 2 inches long, and 1 inch high.

A. 2 square inches

B. 76 square inches

C. 24 square inches

D. 10 square inches

2. Find the volume of the right circular cylinder shown.

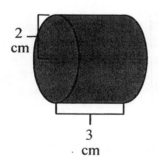

A. 12π cu. cm

B. 18π cu. cm

C. 12π sq. cm

D. 18π sq. cm

3. Find the volume of a rectangular solid measuring 1 meter long, 4 meters wide, and 2 meters high.

A. 8 sq. m

B. 28 sq. m

C. 8 cu. m

D. 28 cu. m

SECTION 10.5

4. Find the volume of the right circular cone shown below.

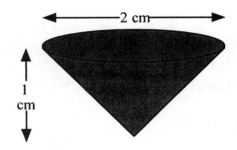

A. π cubic cm

B. $\dfrac{1}{3}\pi$ cubic cm

C. $\dfrac{2}{3}\pi$ cubic cm

D. $\dfrac{4}{3}\pi$ cubic cm

5. Find the volume of a sphere whose radius is 10 inches.

A. $\dfrac{4}{3}\pi$ cubic inches

B. $\dfrac{1000}{3}\pi$ cubic inches

C. $\dfrac{4000}{3}\pi$ cubic inches

D. $\dfrac{400}{3}\pi$ cubic inches

6. Find the volume of a right circular cone whose height is 6 feet and radius is 3 feet.

A. 18π cubic yards

B. 36π cubic yards

C. $\dfrac{2}{3}\pi$ cubic yards

D. 6π cubic yards

7. Find the volume of a right circular cylinder with a diameter of 8 feet and height of 10 feet.

A. 160π cubic feet

B. 640π cubic feet

C. 800π cubic feet

D. 200π cubic feet

8. Find the surface area of the rectangular solid shown.

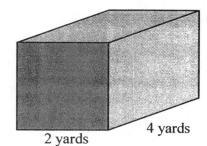

2 yards

4 yards

2 yards

A. 432 cubic feet

B. 432 square feet

C. 360 cubic feet

D. 360 square feet

9. Find the volume of a sphere whose radius is 3000 m.

A. 36π cubic km

B. $36,000,000,000\pi$ cubic km

C. $36,000,000\pi$ cubic km

D. $36,000\pi$ cubic km

10. Find the volume of a rectangular solid measuring 2 yards long, 3 yards wide and 3 feet high.

A. 6 cubic feet

B. 162 cubic feet

C. 172 cubic feet

D. 18 cubic feet

CLAST SKILL IV.B.1 *The student solves real-world problems involving perimeters, areas, and volumes of geometric figures.*

See page 57 of this book to learn about this CLAST skill.

11. A rectangular gravel walkway will be 3 feet wide, 300 feet long and the gravel is packed 9 inches deep. Assuming that gravel costs $8 per cubic yard, what is the cost of the gravel for this walkway?

A. $25

B. $200

C. $21,600

D. $5,400

12. A building caretaker finds that a cylindrical container with a radius of 6 inches and height of 18 inches will hold enough floor wax to last for 9 months. Another cylindrical container of floor wax has a radius of one foot and height of 3 feet. How long will the second container of wax last?

A. 18 months   B. 1.5 months   C. 3 months   D. 72 months

13. How many toy building blocks, each measuring 2 inches by 2 inches by 2 inches, will fit into a shoebox that measures 6 inches by 6 inches by 1 foot?

A. 54   B. 27   C. 4.5   D. 45

14. A cone-shaped vessel with a height of 1/2 foot and radius of 2 inches is used to scoop sand out of a cylindrical bucket that has a height of one foot and radius of 1/2 foot. How many of the cone-shaped scoops are required to hold the same amount of sand as the bucket?

A. 12   B. 18   C. 36   D. 54

15. A rectangular tank measuring 9 feet wide, 12 feet long and 6 feet high is filled with a liquid that weighs 1000 pounds per cubic yard. What is the total weight of the liquid in this tank?

A.   1,944,000 pounds

B.   24,000 pounds

C.   72,000 pounds

D.   .024 pounds

16. A portable air cleaner can treat 12 cubic yards of air in 1 hour. How long will it take to treat the air in a room measuring 12 feet long, 24 feet wide and 9 feet high?

A. 216 hours

B. 72 hours

C. 24 hours

D. 8 hours

CLAST SKILL III.B.1 *The student infers formulas for measuring geometric figures.*

See page 49 of this book to learn about this CLAST skill.

17. Study the pyramids shown in the figure below. In each figure, L represents the lengths of both sides of the square base as well as the height of the pyramid, and V represents the volume.

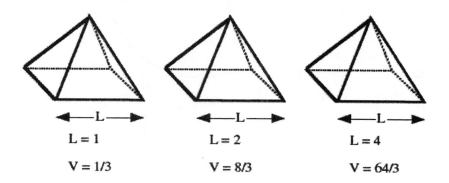

L = 1

V = 1/3

L = 2

V = 8/3

L = 4

V = 64/3

For a similar pyramid, find V if L = 10.

A. 100/3

B. 125/3

C. 1,000/3

D. 1,250/3

18. Study the figures below, each of which shows a sphere and a cross-sectional circle. In each figure, A is the area of the cross-sectional circle and V is the volume of the sphere.

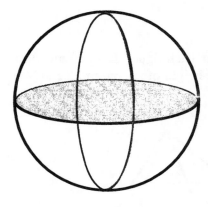

$$A = 9\pi$$

$$V = 36\pi$$

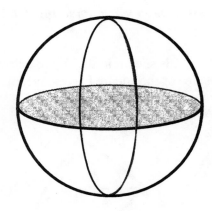

$$A = 25\pi$$

$$V = \frac{500\pi}{3}$$

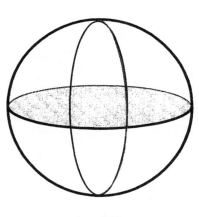

$$A = 36\pi$$

$$V = 288\pi$$

Find the volume of a similarly constructed figure if $A = 81\pi$.

A. 324π B. 972π

C. 108π D. 1440π

19. Study the wedge-shaped figures (prisms) shown below. In each figure, V represents volume.

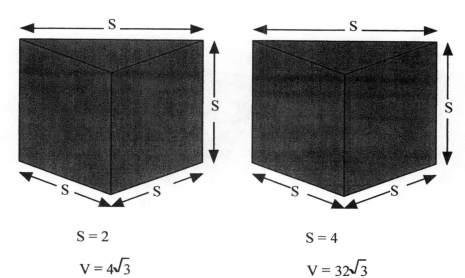

$$S = 2$$

$$V = 4\sqrt{3}$$

$$S = 4$$

$$V = 32\sqrt{3}$$

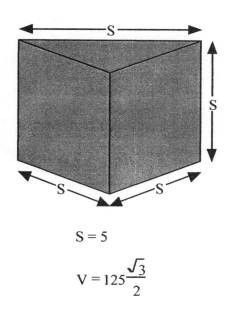

$$S = 5$$

$$V = 125\frac{\sqrt{3}}{2}$$

For a similar figure, find V if S = 1.

A. $3\dfrac{\sqrt{3}}{2}$ B. $2\sqrt{3}$ C. $\sqrt{3}$ D. $\dfrac{\sqrt{3}}{2}$

SECTION 10.5

20. Study the figures below, each of which shows a right circular cone inscribed within a right circular cylinder. In each figure, v represents the volume of the cone, while V represents the volume of the solid that remains when the cone has been extracted from the cylinder.

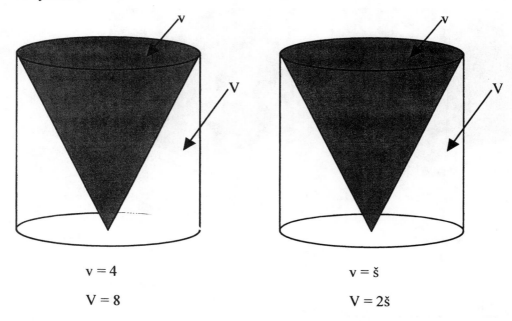

$$v = 4$$

$$V = 8$$

$$v = \check{s}$$

$$V = 2\check{s}$$

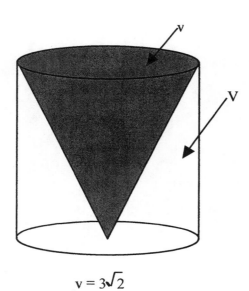

$$v = 3\sqrt{2}$$

$$V = 6\sqrt{2}$$

For a similarly constructed figure, find V if v = 1.5.

A. $1.5\sqrt{2}$ B. 1.5π C. 3   D. 6

21. Study the solid figures shown below. In each figure, the top surface is parallel to and congruent to the bottom surface (which is hidden from view) and perpendicular to the other surfaces. In each case, A represents the area of the top surface, h represents the height, and V represents the volume.

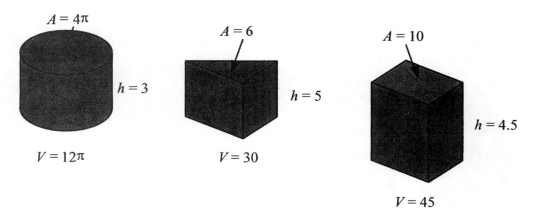

Now find the volume of this similarly constructed figure:

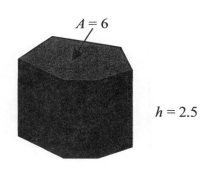

$V = ?$

A. 8.5

B. 15

C. 90

D. 38.5

SECTION 10.5

CLAST SKILL III.B.2 The *student identifies applicable formulas for computing measures of geometric figures.*

See page 54 of this book to learn about this CLAST skill.

22. Study the figure below, showing a right circular cone joined to a right circular cylinder, and then select the formula for calculating the volume (V) of the figure. The radius of the circular base is r.

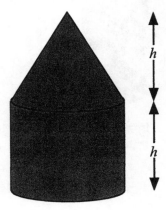

A. $V = \dfrac{4}{3}\pi r^3$

B. $V = 2\pi r^2 h$

C. $V = \dfrac{2}{3}\pi r^2 h$

D. $V = \dfrac{4}{3}\pi r^2 h$

23. Study the figure below, which shows two rectangular solids joined together, and then select the formula for computing the volume (V) of the figure.

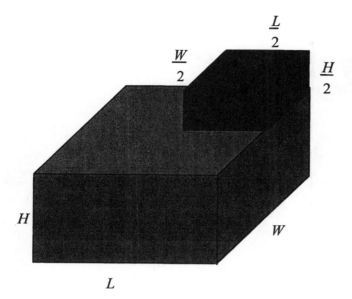

A. $V = L \cdot W \cdot H + \dfrac{L}{2} \cdot \dfrac{W}{2} \cdot \dfrac{H}{2}$

B. $V = (L + W + H)\left(\dfrac{L}{2} + \dfrac{W}{2} + \dfrac{H}{2}\right)$

C. $V = L \cdot W \cdot H - \dfrac{L}{2} \cdot \dfrac{W}{2} \cdot \dfrac{H}{2}$

D. $V = \dfrac{L + W + H}{\dfrac{L}{2} + \dfrac{W}{2} + \dfrac{H}{2}}$

SECTION 10.5

24. Study the figure below that shows two cones joined together at their circular bases.

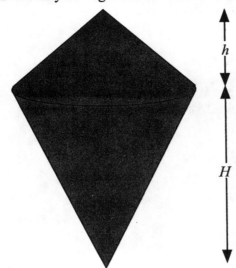

Select the formula for computing the volume (V) of the figure.

A. $V = \dfrac{1}{3}\pi r^2 hH$

B. $V = \dfrac{2}{3}\pi r^2 hH$

C. $V = \dfrac{1}{3}\pi r^2 h + \dfrac{1}{3}\pi r^2 H$

D. $V = \dfrac{2}{3}\pi r^2 h + \dfrac{2}{3}\pi r^2 H$

25. Study the figure below, showing how two congruent wedge-shaped figures (prisms) can be combined to form a rectangular solid and then select the formula for calculating the volume (V) of one of the wedge-shaped prisms.

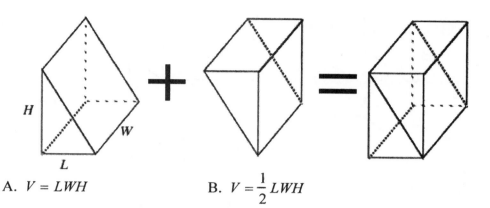

A. $V = LWH$ B. $V = \dfrac{1}{2}LWH$

C. $V = \dfrac{1}{2}L \cdot \dfrac{1}{2}W \cdot \dfrac{1}{2}H$ D. $V = \dfrac{1}{2}L + \dfrac{1}{2}W + \dfrac{1}{2}H$

26. Study the figure below, consisting of two right circular cylinders joined together. Assume that the radius of the larger cylinder is twice the radius of the smaller cylinder. Select the formula for the volume (V) of the figure.

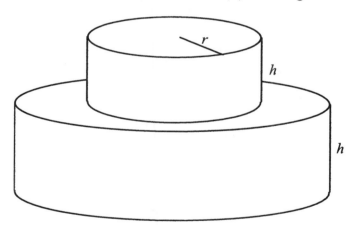

A. $V = \dfrac{1}{2}\pi r^2 h + \pi r^2 h$ B. $V = \pi r^2 h + 2\pi r^2 h$

C. $V = \pi r^2 h + 4\pi r^2 h$ D. $V = \pi r^2 h + 3\pi r^2 h$

27. Study the figure below, showing two cubes combined to form a larger rectangular solid.

SECTION 10.5

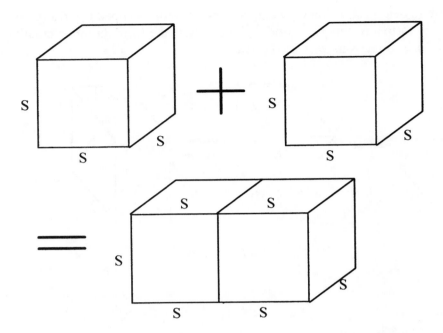

Select the formula for computing the surface area (SA) of the larger solid.

A. $SA = 6S^2$

B. $SA = 12S^2$

C. $SA = 10S^2$

D. $SA = 5S^2$

CLAST Exercises for Section 11.1 of *Thinking Mathematically.*

CLAST SKILL I.D.3 *The student will use the fundamental counting principle.*

See page 108 of this book to learn about this CLAST skill.

1. As a result of changing their telephone long distance service providers, Sandy, Sally, Ben and Billy will receive rewards. Each of them may choose a portable stereo, a television, or a cordless telephone. How many different outcomes are possible?

A. 12   B. 81   C. 64   D. 7

2. A certain model of computer may be purchased with any combination of the following optional items: i. extended warranty; ii. modem; iii. printer; iv. scanner; v. digital camera. How many different combinations of optional items are there?

A. 32   B. 25   C. 120   D. 3125

3. *Das Boot*, the new submarine sandwich shop, is offering a lunch special. The customer must choose either a 6" or 12" sandwich, either white bread or whole wheat bread, and either beef, ham or turkey breast. How many different outcomes are possible?

A. 9   B. 8   C. 12   D. 7

4. There are three motorists who are going to wash their cars at a coin-operated car wash. Each of them may choose the Deluxe service, the Super service, the Super-Deluxe service or the Extra-Super-Deluxe service. How many different outcomes are possible?

A. 64   B. 7   C. 81   D. 12

5. A street vendor is selling ice cream cones. The following flavors are offered: chocolate, vanilla, strawberry, mint. The following types of cone are available: sugar cone, waffle cone. The purchaser must also decide whether to order one scoop, two scoops or three scoops. Anne is going to order an ice cream cone, but she knows that she won't order chocolate. How many outcomes are possible, assuming that regardless of how many scoops are chosen they will all be the same flavor?

A. 36   B. 8   C. 6   D. 18

SECTION 11.1

6. A mail carrier is approaching the end of his daily route. There are five mailboxes remaining to be served, and the mail carrier has only two letters left in his mailbag. Assuming each of the two letters is destined for one of the five remaining mailboxes, in how many ways is it possible for the two letters to be distributed among the five mailboxes?

A. 10   B. 25   C. 1,024   D. 100

7. Jamie is going to wrap a package containing a birthday gift. She may choose blue, green or purple wrapping paper, either red or pink ribbon, and either a red or pink bow. How many outcomes are possible, assuming that the ribbon and bow will be the same color?

A. 7   B. 8   C. 12   D. 6

8. Andy, Will and Frank have to perform three chores today: wash laundry, scrub the floor, and drive to the grocery. They will each do one chore, but Andy cannot drive to the store because his license is suspended. In how many ways may the three chores be assigned to the three people?

A. 4   B. 6   C. 5   D. 2

9. The film club is going to select a musical, a drama and a comedy for a film festival. The musicals from which they can choose are *Oliver, Hair, Fiddler on the Roof* and *My Fair Lady*. The dramas are *The Godfather, On the Waterfront* and *A Streetcar Named Desire*. The comedies are *Horse Feathers, It's a Mad, Mad, Mad, Mad World* and *Some Like It Hot*. How many outcomes are possible if they have already decided that they won't choose *Horse Feathers* and they will choose *Oliver*.

A. 18   B. 9   C. 3   D. 6

10. Whenever he has to create a Personal Identification Number (PIN) for a bank or credit card account, Harlan always picks an even number whose digits are chosen from this set: 3, 4, 5, 6, 7. How many different 4-digit PINs are possible under this scheme?

A. 250   B. 64   C. 686   D. 620

CLAST Exercises for Section 11.2 of *Thinking Mathematically.*

CLAST SKILL I.D.3 *The student will use the fundamental counting principle.*

See page 108 of this book to learn about this CLAST skill.

1. The English Department is going to hire three different part-time instructors, one each to teach one section of ENC1101, ENC1102 and ENC2211 respectively. There are six instructors from whom to choose, each of whom is equally well qualified to teach any of the three courses. In how many ways is it possible to make the assignments?

A. 216   B. 720   C. 120   D. 18

2. On a youth league baseball team there are five players who share playing time at the three infield positions of second base, third base and shortstop. The five players are similarly talented and are all roughly equally suited to playing any of those three positions. If the manager randomly chooses three of the players and assigns one to each of the three positions, how many outcomes are possible?

A. 32   B. 60   C. 120   D. 125

3. There are six applicants for the position of sales manager. The members of the selection committee will rank the six applicants from "best qualified" to "least qualified." How many outcomes are possible?

A. 720   B. 120   C. 36   D. 24

4. There are four men and six women on the governing board of the local chapter of the Purple Party. From amongst themselves they will select a group of four people to attend the Purple Party national convention. One delegate and one alternate will be chosen from among the men, and one delegate and one alternate will be chosen from among the women. How many four-person groups are possible?

A. 42   B. 360   C. 5,040   D. 34

5. The Personal Identification Number (PIN) for Amanda's debit card is a four-digit number formed from the digits 2, 5, 6, and 8 with no repeated digits. How many different PINs are possible?

A. 24   B. 256   C. 16   D. 64

SECTION 11.2

6. In European History class today there are three students who have volunteered to present their oral reports. Before class they must decide who will go first, who will go second and who will go last. How many arrangements are possible?

A. 9   B. 6   C. 64   D. 36

7. A disk jockey has promised her friends to play four specific songs before the end of her shift, but she realizes that she only has enough time to play three songs. How many different three-song arrangements are possible?

A. 12   B. 7   C. 24   D. 84

8. Six volunteers are participating in an experiment testing the value of five different allergy medications. Five of them will each receive a different medicine, and the sixth will receive a placebo. In how many ways may the substances be distributed among the six people?

A. 30   B. 120   C. 720   D. 144

9. A candidate for the state assembly is going to schedule visits to five cities in his district. First he will visit three cities having populations less than 10,000, and then two cities having populations greater than 10,000. He will choose and arrange the visits from a list of six cities having populations less than 10,000 and four cities having populations greater than 10,000. In how many ways can he arrange the five visits?

A. 720   B. 55,440   C. 132   D. 1,440

10. A homeowner is going to have the exterior of her house painted. She has a collection of five colors from which she will choose one color for the exterior walls and a second color for the trim. How many color combinations are possible?

A. 25   B. 10   C. 20   D. 120

CLAST Exercises for Section 11.3 of *Thinking Mathematically.*

CLAST SKILL I.D.3 *The student will use the fundamental counting principle.*

See page 108 of this book to learn about this CLAST skill.

1. A restaurant lunch special allows the customer to choose two vegetables from this list: okra, corn, peas, carrots, and squash. How many outcomes are possible if the customer chooses two different vegetables?

A. 10   B. 20   C. 25   D. 32

2. There are six employees in the stock room at an appliance retail store. The manager will choose three of them to deliver a refrigerator. How many three-person groups are possible?

A. 18   B. 216   C. 120   D. 20

3. Bernie has three dress shirts, two ties and two jackets. He needs to select a dress shirt, a tie and a jacket for work today. How many outcomes are possible?

A. 35   B. 12   C. 210   D. 7

4. Curt has four flannel shirts. He is going to choose two of them to take on a camping trip. How many outcomes are possible?

A. 12   B. 6   C. 8   D. 16

5. In his kitchen, Chef Jean has five brands of Louisiana hot sauce. He will choose three of them to mix into his gumbo. How many outcomes are possible?

A.15   B. 60   C. 10   D. 125

6. There are four Democrats and four Republicans on the county commission. From among their group they will choose a committee of two Democrats and two Republicans to examine a proposal to purchase land for a new county park. How many four-person groups are possible?

A. 12   B. 36   C. 70   D. 1680

SECTION 11.3

7. In the Mathematics Department there are four female professors and six male professors. Three female professors will be chosen to serve as mentors for a special program designed to encourage female students to pursue careers in mathematics. In how many ways may the selection be made?

A. 12   B. 64   C. 4   D. 120

8. An office employs six customer service representatives. Each day, two of them are randomly selected and their customer interactions are monitored for the purposes of improving customer relations. In how many ways may the selection be made?

A. 30   B. 12   C. 8   D. 15

9. James is going to form a four-letter password for his Internet account. The four letters will be chosen from this set: a, d, h, n, p, w. How many four-letter passwords are possible, assuming that the password will have no repeated letters?

A. 15   B. 360   C. 24   D. 10

10. A zoo has three male bears and five female bears. One male bear and two female bears will be selected for an animal exchange program with another zoo. How many three-bear collections are possible?

A. 56   B. 336   C. 30   D. 60

11. A youth group is going to occupy five campsites at a campground. There are six campsites from which to choose. In how many ways may the selection be made?

A. 5   B. 6   C. 30   D. 720

12. Billy and Sally have driven to work in separate cars. When they arrive, there are five empty spaces in the parking lot. They will each choose a parking space. How many outcomes are possible?

A. 20   B. 10   C. 25   D. 32

13. Four delegates meet to negotiate a contract between the auto workers and the manufacturing company. Prior to taking seats, each person shakes hands with each of the others. How many handshakes occur?

A. 8   B. 4   C. 12   D. 6

CLAST Exercises for Section 11.4 of *Thinking Mathematically.*

CLAST SKILL IV.D.2 *The student solves real-world problems involving probabilities.*

See page 135 of this book to learn about this CLAST skill.

1. A security company monitors alarm calls from automatic security systems installed in private residences. The nature of such alarm calls is summarized in the table below.

reason for alarm call	percent of alarm calls
intruder in house	7%
intruder on premises but not in house	11%
smoke detected in house	18%
animal triggered alarm	14%
user error	12%
equipment malfunction	12%
unknown cause	26%

Out of 600 alarm calls, how many would be expected to be caused by "equipment malfunction?"

A. 12   B. 62   C. 72   D. 50

2. A security company monitors alarm calls from automatic security systems installed in private residences. The nature of such alarm calls is summarized in the table below.

reason for alarm call	percent of alarm calls
intruder in house	7%
intruder on premises but not in house	11%
smoke detected in house	18%
animal triggered alarm	14%
user error	12%
equipment malfunction	12%
unknown cause	26%

What is the probability that the reason for an alarm call is "smoke detected in house?"

A. .018   B. .18   C. .82   D. .22

SECTION 11.4

3. The pie chart below shows the classification according to academic major of the students in a liberal arts math class. None of the students have double majors.

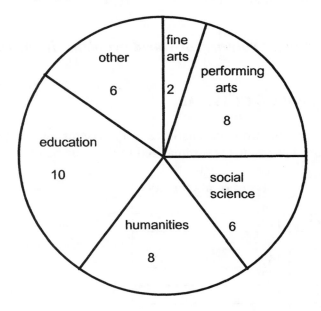

If one student is selected, what is the probability that he or she is an education major?

A. 10/100   B. 10/30   C. 30/10   D. 10/40

4. A count of cars in a shopping center parking lot revealed the following information about the country in which the car's manufacturer was based.

country	number of cars
United States	56
Japan	40
Germany	16
South Korea	18
United Kingdom	2
France	2
Italy	1
Sweden	14
country no longer exists	1

If one car is selected, what is the probability that its manufacturer was based in South Korea?

A. .18   B. .09   C. .12   D. .9

5. The table below summarizes characteristics of applicants to a law school.

		Household Income		
		less than $35,000	$35,000 or more	totals
Sex	male	32%	14%	46%
	female	34%	20%	54%
	totals	66%	34%	100%

What is the probability that a randomly selected applicant is male?

A. 46/100   B. 32/36   C. 32/46   D. 46/54

6. The table below summarizes characteristics of applicants to a law school.

		Household Income		
		less than $35,000	$35,000 or more	totals
Sex	male	32%	14%	46%
	female	34%	20%	54%
	totals	66%	34%	100%

Out of 1200 applicants, how many would be expected to have household incomes less than $35,000?

A. 384   B. 408   C. 792   D. 660

7. The table below classifies a group of students according to the number of hours per week spent using the Internet.

hours per week	percent
less than 2.00	16%
2.00 - 6.99	38%
7.00 – 13.99	32%
14.00 – 20.99	12%
21 or more	2%

Find the probability that a randomly chosen student uses the Internet for 21 or more hours per week.

A. .2   B. .02   C. .98   D. .098

SECTION 11.4

CLAST SKILL II.D.3 *The student will identify the probability of a specified outcome in an experiment.*

See page 117 of this book to learn about this CLAST skill.

8. A manufacturer employs 42 women and 35 men. If one employee is randomly selected, what is the probability that he or she is a man?
A. 42/35   B. 35/42   C. 35/77   D. 35/100

9. Students at a certain university having a "party school" reputation were surveyed in the spring of 2000. On the subject of alcohol consumption, 54% admitted to binge drinking, 39% reported being drunk three or more times in the previous month, and 28% reported having missed classes as a result of excessive drinking. What is the probability that a randomly selected student has missed classes because of excessive drinking?
A. .39   B. .54   C. .30   D. .28

10. A retailer surveyed customers and obtained the following information.

40% own a computer.   35% own a cellular phone.
25% own both a computer and a cellular phone.

What is the probability that a randomly selected customer owns both a computer and a cellular phone?
A. .25   B. .75   C. 1.00   D. 100

11. A retailer surveyed customers and obtained the following information.

40% own a computer.   35% own a cellular phone.
25% own both a computer and a cellular phone.

Out of 50 customers, how many would be expected to own a computer.
A. 40   B. 20   C. 35   D. 15

12. In a group of 30 preschool children, 12 were able to print their names and 6 were able to print their telephone numbers. Find the probability that a randomly selected child is able to print his or her name.
A. .4   B. .12   C. .012   D. .5

13. Researchers expect that 74% of adults will achieve scores greater than 90 on a test commonly used to measure IQ. If one person is randomly selected, find the probability that he or she will achieve a score greater than 90.
A. .37   B. .90   C. .74   D. .10

CLAST Exercises for Section 11.5 of *Thinking Mathematically.*

CLAST SKILL II.D.3 *The student will identify the probability of a specified outcome in an experiment.*

See page 117 of this book to learn about this CLAST skill.

1. Among the nine starting players on a baseball team, two are left-handed. If two players of the nine are randomly selected, what is the probability that both are left-handed?

A. 2/9   B. 1/9   C. 1/36   D. 1/18

2. Charlie, Mike, Jenny, Sara, Allan, and Fran are going to a concert. They will draw lots to choose two designated drivers. Find the probability that neither Allan nor Fran will be the designated drivers.

A. 2/5   B. 1/3   C. 1/12   D. 1/6

3. Lawrence's Internet account password consists of three characters, with no repeated characters. If the characters are chosen from the set {a, b, c, d}, what is the probability that the password is "dba?"

A. 1/3   B. 1/24   C. 1/12   D. 1/4

4. Jeremy is taking an 8-question true/false quiz. He knows the answers to three of the questions, but has no clue on the other five questions. If he just guesses at those five questions, what is the probability that his score on the quiz will be 100%?

A. 1/32   B. 1/8   C. 5/8   D. 1/5

5. In her pocket, Donna has three $1 bills, two $5 bills, one $10 bill and one $20 bill. If she reaches into her pocket and grabs two bills, what is the probability that their monetary sum will be $30?

A. 2/7   B. 30/43   C. 1/21   D. 1/14

6. There are six houses on the block where Millie lives. Three of the houses have building code violations. If two of the houses are inspected, what is the probability that they both have building code violations?

A. 1/3   B. 2/3   C. 1.6   D. 1/5

CLAST Exercises for Section 11.6 of *Thinking Mathematically.*

CLAST SKILL IV.D.2 *The student solves real-world problems involving probabilities.*

See page 135 of this book to learn about this CLAST skill.

1. A security company monitors alarm calls from automatic security systems installed in private residences. The nature of such alarm calls is summarized in the table below.

reason for alarm call	percent of alarm calls
intruder in house	7%
intruder on premises but not in house	11%
smoke detected in house	18%
animal triggered alarm	14%
user error	12%
equipment malfunction	12%
unknown cause	26%

What is the probability that the reason for an alarm call was "intruder in house" or "intruder on premises but not in house?"

A. .77   B. .077   C. .91   D. .18

2. The table below contains information from the final round of the 1998 World Cup soccer championship, showing the distribution of games according to the total number of goals scored by both teams (goals scored on penalty kicks are not included).

Total number of goals scored	Percent of games
0	.0625
1	.1875
2	.0625
3	.4375
4	.0625
5	.1875

If one game is randomly selected, what is the probability that fewer than three goals were scored?

A. .75   B. .3125   C. .4375   D. .25

3. Referring to the data in the previous exercise, if 16 games were played, in how many games would we expect that fewer than 5 goals were scored?

A. 3   B. 1   C. 5   D. 13

4. The table below summarizes information collected from incoming college freshmen who were asked to classify their own political beliefs (source: The *American Freshman: National Norms for Fall 1997*).

political classification	percent of freshmen
"conservative" or "far right"	20.8%
"middle of the road"	54.8%
"liberal" or "far left"	24.4%

What is the probability that an incoming freshman's political beliefs are not "middle of the road?"

A. .552   B. .542   C. .244   D. .452

5. The pie chart below shows the estimated distribution according to age (in years) of the population of the United States in 1995. *(Source: Bureau of the Census, U.S. Department of Commerce.)*

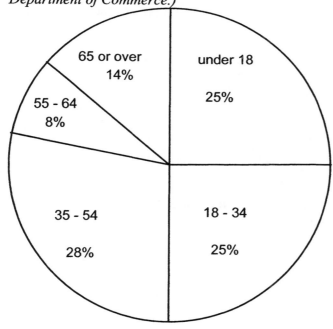

What is the probability that a randomly selected U.S. citizen's age is in the 18 – 54 year range?

A. .25   B. .50   C. .53   D. .47

SECTION 11.6

6. The table below presents information about U.S. firearms deaths in 1995 (*source: National Safety Council.*)

Firearms Deaths 1995				
		age of victim		
		under 15	15 or older	totals
cause of	accidental	0.5%	2.9%	3.4%
death	not accidental	1.9%	94.7%	96.6%
	totals	2.4%	97.6%	100%

What is the probability that a firearm fatality was accidental or involved a victim whose age was under 15?

A. .005

B. .05

C. .048

D. .053

7. The table below gives information about fatal occupational injuries in *1997 (source: Bureau of Labor Statistics, U.S. Department of Labor).*

Cause	Percent
Transportation Incidents	42%
Assaults and Violent Acts	18%
Falls	11%
Contact with Objects and Equipment	17%
Exposure to Harmful Substances	9%
Fires and Explosions	3%

Find the probability that a fatal occupational accident was neither caused by "exposure to harmful substances" nor "fires and explosions."

A. .12

B. .27

C. .88

D. .73

8. The graph below shows the distribution, according to car size and type, of cars sold in the U.S. in 1997 *(source: American Automobile Manufacturers Association).*

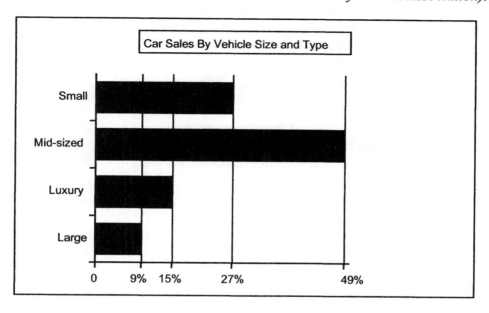

If one of these cars is randomly selected, find the probability that it is neither "Small" nor "Luxury."

A. .73   B. .42   C. .58   D. .85

CLAST SKILL II.D.3   *The student will identify the probability of a specified outcome in an experiment.*

See page 117 of this book to learn about this CLAST skill.

9. According to the American Medical Association, in 1997, 22% of physicians in the U.S. were female, 18% were under 35 years old, and 6% were females who were under 35 years old. If one of these physicians is randomly selected, find the probability that he or she is female or under 35 years old.

A. .40   B. .34   C. .46   D. .06

10. According to estimates from the Social Security Administration, 72% of Social Security beneficiaries are at least 65 years old, 62% are retired workers, and 50% are retired workers who are at least 65 years old. Find the probability that a randomly selected Social Security beneficiary is a retired worker who isn't at least 65 years old.

A. 3/25   B. 1/2   C. 11/50   D. 4/25

11. After the November 1998 elections the governors of 17 states were Democrats, the governors of three states were women, and the governor of one state was a Democrat woman. Based on those numbers, if one state governor was randomly selected, find the probability that the governor was not a Democrat and was not a woman.

A. 47/50   B. 31/50   C. 28/50   D. 29/50

12. After the November 1998 elections the governors of 17 states were Democrats, the governors of three states were women, and the governor of one state was a Democrat woman. Based on those numbers, if one state governor was randomly selected, find the probability that the governor was not a woman.

A. 1/50   B. 3/50   C. 24/25   D. 47/50

13. Among the players chosen in the first round of the 1998 National Basketball Association draft, 11 played guard, 16 played forward, 3 played both guard and forward, while 5 played neither of those positions. If one of these players is randomly selected, find the probability that he played guard or forward.

A. 30/35   B. 27/35   C. 27/29   D. 24/29

14. According to estimates from the American Cancer Society, among the U.S. citizens who died from cancer in 1998, 52% were men, 28% died from lung cancer, and 16% were men who died from lung cancer. If one of these cases is randomly selected, find the probability that the victim was a man who didn't die from lung cancer.

A. .36   B. .48   C. .12   D. .72

15. According to data from the U.S. Census Bureau (March 1998 Current Population Survey) among U.S. residents whose educational attainment is "bachelor's degree or higher" 8.2% do not have health insurance coverage. If a U.S. resident whose educational attainment is bachelor's degree or higher is randomly selected, find the probability that he or she has health insurance coverage.

A. .82   B. .082   C. .18   D. .918

CLAST Exercises for Section 11.7 of *Thinking Mathematically*.

CLAST SKILL II.D.3 *The student will identify the probability of a specified outcome in an experiment.*

See page 117 of this book to learn about this CLAST skill.

1. According to the Bureau of the Census (March 1997), 11% of people employed in "farming, forestry [or] fishing" have educational attainment of at least a bachelor's degree. If we randomly select two people who are employed in "farming, forestry [or] fishing" what is the probability that both of them have educational attainment of at least a bachelor's degree?

A. .0121   B. .22   C. .11   D. .89

2. Among the top five picks in the 1998 National Football League draft, there were two quarterbacks, one running back, one defensive lineman and one defensive back. If we randomly choose two people from this population, find the probability that both are quarterbacks.

A. 2/5   B. 1/10   C. 1/5   D. 4/25

3. According to data from the National Safety Council, among people who died in 1997 as a result of home accidents, 32% died as a result of falls. If we randomly select two cases involving people who died from home accidents in 1997, what is the probability that neither of them died as a result of a fall?

A. .1024   B. .68   C. .64   D. .4624

4. According to the Automobile Manufacturers Association, in 1997, 15% of new cars sold in the U.S. were luxury cars and 10% of luxury cars were colored white. Find the probability that a new car sold in the U.S. in 1997 was a white luxury car.

A. .05   B. .25   C. .015   D. .667

5. According to the U.S. Census Bureau (1997), 20% of children under the age of 18 live in poverty. According to this data, if two children under age 18 are randomly selected what is the probability that at least one of them lives in poverty?

A. .36   B. .2   C. .02   D. .04

6. According to ABC News, in Los Angeles 62% of police car chases begin with traffic violations, and 50% of those car chases end with crashes. What is the probability that a car chase begins with a traffic violation and doesn't end with a crash?

A. .132   B. .31   C. .0132   D. .50

7. Mark, Sammy and Chuck are going bowling. They will randomly decide who bowls first, who bowls second and who bowls third. What is the probability that Chuck bowls first, Sammy bowls second and mark bowls third?

A. 1/6   B. 1/3   C. 1/8   D. 1/27

8. According to the U.S. Census Bureau (1997), 15% of U.S. residents under the age of 18 are not covered by health insurance. If two residents under the age of 18 are randomly selected, find the probability that neither of them is covered by health insurance.

A. .15   B. .7225   C. .0225   D. .17

9. A manufacturer produces frozen burritos that are supposed to weigh 4 ounces each. Due to random errors in the manufacturing process 25% of the burritos weigh more than 4.2 ounces. If two burritos are randomly selected, find the probability that at least one of them weighs more than 4.2 ounces.

A. .50   B. .4375   C. .0625   D. .25

CLAST SKILL IV.D.2   *The student solves real-world problems involving probabilities.*

See page 135 of this book to learn about this CLAST skill.

10. The table below shows the distribution according to age of the children in a pre-school program.

age	percent
3	60%
4	30%
5	10%

Find the probability that a child is three years old, given that he or she is less than five years old.

A. 3/5   B. 1/2   C. 2/3   D. 1/3

11. A number of automobile shoppers were surveyed regarding their primary interests. The results are summarized in the table below.

	want new vehicle	want used vehicle	totals
want car	30%	40%	70%
want truck	10%	20%	30%
totals	40%	60%	100%

If two people are randomly selected, find the probability that at least one of them wants a car that is new.

A. .6   B. .3   C. .51   D. .49

12. A number of automobile shoppers were surveyed regarding their primary interests. The results are summarized in the table below.

	want new vehicle	want used vehicle	totals
want car	30%	40%	70%
want truck	10%	20%	30%
totals	40%	60%	100%

If one person is randomly selected, find the probability that he/she wants a used vehicle, given that he/she wants a car.

A. 2/5   B. 3/5   C. 2/3   D. 4/7

13. The table below shows the distribution of tourists visiting the U.S., according to their native land. *(Source: Tourism Industries, International Trade Administration, U.S. Department of Commerce.)*

Country	percent
Canada	33%
Mexico	18%
Japan	11%
United Kingdom	7%
Germany	4%
Other	27%

If two tourists are randomly selected find the probability that they are both from Germany.

A. .8   B. .0016   C. .08   D. .0784

14. A group of high school freshmen and sophomores were given an IQ test. The distribution of students according to IQ score is shown in the table below.

IQ score	percent of students
less than 70	1%
71 – 85	15%
86 – 100	36%
101 – 115	33%
116 – 130	13%
131 or more	2%

Find the probability that a randomly selected student's score is more than 85, given that it is less than 116.

A. 69/85   B. 84/85   C. 84/100   D. 69/100

15. Referring to the data in the previous problem, if two students are randomly selected, find the probability that at least one of them has an IQ in the 71 – 85 range.

A. .0225   B. .225   C. .3   D. .2775

16. The table below contains information from the final round of the 1998 World Cup soccer championship, showing the distribution of games according to the total number of goals scored by both teams (goals scored on penalty kicks are not included).

Total number of goals scored	Proportion of games
0	.0625
1	.1875
2	.0625
3	.4375
4	.0625
5	.1875

If one game is randomly selected, what is the probability that fewer than 4 goals were scored, given that more than one goal was scored?

A. 50/100

B. 75/100

C. 50/75

D. 4375/5000

314

17. The pie chart below shows the estimated distribution according to age (in years) of the population of the United States in 1995. *(Source: Bureau of the Census, U.S. Department of Commerce.)*

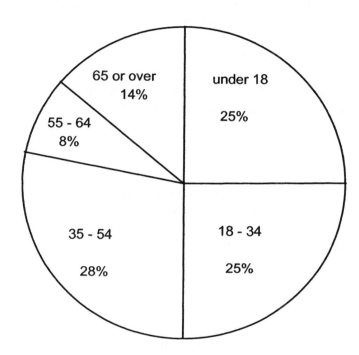

What is the probability that a randomly selected U.S. resident's age is 65 or over, given that it is 55 or over?

A. 8/14   B. 22/100   C. 14/22   D. 14/100

18. Referring to the data in the previous problem, if two U.S. residents are randomly selected what is the probability that neither of them is in the 18 – 34 year age range?

A. .0625   B. .5625   C. .9375   D. .4375

CLAST Exercises for Section 12.1 of *Thinking Mathematically.*

CLAST SKILL II.D.2 *The student will choose the most appropriate procedure for selecting an unbiased sample from a target population.*

See page 115 of this book to learn about this CLAST skill.

1. A jewelry marketer wants to survey a sample of Davis County high school sophomores in order to determine preferences for class rings. Which of the following procedures would be most appropriate for obtaining a statistically unbiased sample?

A. Mail surveys to a random sample of Davis County households.

B. Randomly select an assortment of classes for sophomores at each Davis County high school and distribute surveys to the students in those classes.

C. Randomly select an assortment of Davis County high school teachers and distribute surveys to them.

D. Distribute surveys to all Davis County sophomores on the honor roll.

2. A public radio station wants to survey its contributors to determine their programming interests. Which of the following procedures would be most appropriate for obtaining a statistically unbiased sample?

A. Survey the 100 people who made the largest contributions to the station.

B. Mail surveys to 100 people randomly chosen from the local voter registration rolls.

C. Survey 100 people randomly chosen from the list of people who send e-mail to the station.

D. Survey 100 people randomly chosen from the list of people who have contributed to the station within the past year.

3. The Executive Committee of the Gaspar County Republican Party wants to survey local registered Republican voters in order to determine which issues should be stressed in local elections. Which of the following procedures would be most appropriate for obtaining a statistically unbiased sample?

A. Select the first 50 people listed in the Gaspar County voter registration rolls.

B. Randomly select 50 people from the Gaspar County voter registration rolls.

C. Randomly select 50 registered Republicans from the Gaspar County voter registration rolls.

D. Select the first 50 Republicans from the Gaspar County voter registration rolls.

4. The Neptune Automobile Company wants to survey recent Neptune purchasers in order to determine the level of customer satisfaction. Which of the following procedures would be most appropriate for obtaining a statistically unbiased sample?

A. Invite Neptune owners to stop by their local dealerships for a free oil change in return for filling out a survey.

B. Randomly select 1000 names from a list of recent Neptune purchasers and contact them via the telephone.

C. Randomly contact 1000 people via telephone.

D. Randomly contact 1000 people who each own more than one Neptune.

5. The makers of the Uplook e-mail program want to survey registered users of the Medora e-mail program in order to determine which of the Medora program's features should be incorporated into Uplook. Which of the following procedures would be most appropriate for obtaining a statistically unbiased sample?

A. Randomly select 500 names from the list of Medora's registered users and contact them via telephone.

B. Send surveys via e-mail to all of the subscribers to the America Downline Internet service and ask that only Medora users reply.

C. Randomly select 500 registered Medora users in Miami and contact them via telephone.

D. Randomly select 500 software professionals and contact them via U.S. mail.

SECTION 12.1

6. The City Council wants to survey homeowners to determine the level of support for a new trash collection service. Which of the following procedures would be most appropriate for obtaining a statistically unbiased sample?

A. Drive around neighborhoods and talk to people who have a lot of junk piled up in their yards.

B. Randomly select 50 people from the local telephone directory and contact them via telephone.

C. Randomly select 50 homeowners from the local property tax rolls and contact them via telephone.

D. Place survey forms on bulletin boards at randomly selected shopping centers in the city.

CLAST SKILL III.D.1 *The student infers relations and makes accurate predictions from studying statistical data.*

See page 124 of this book to learn about this CLAST skill.

7. Consider the following plot of run times (in seconds) for Olympic gold medal winners in the men's 400-meter run.

Times for Olympic 400-Meter Run Champion

How many Olympic champions won with times that were faster than 44 seconds?

A. 2   B. 3

C. 4   D. 7

8. The table below shows some rates for airmail, letters, and letter packages sent from the U.S. to other countries (as of 1999). The trend indicated by this data continues for packages up to 64 ounces in weight.

Weight of Package (ounces)	Rate ($)
0.00 – 0.50	0.60
0.51 – 1.00	1.00
1.01 – 1.50	1.40
1.51 – 2.00	1.80
2.01 – 2.50	2.20
2.51 – 3.00	2.60
3.01 – 3.50	3.00
3.51 – 4.00	3.40

Based on the rates shown in this table, what would be the rate for a letter package weighing 4.68 ounces?

A. $4.00   B. $3.80   C. $4.60   D. $4.20

9. The graph below compares air quality in Miami with air quality in Atlanta over the years 1987 – 1996. For each city for each year the graph shows the number of days on which the air quality failed to meet acceptable air quality standards. *(Source: U.S. Environmental Protection Agency, Office of Air Quality Planning and Standards.)*

Unacceptable Air Quality Days by Year

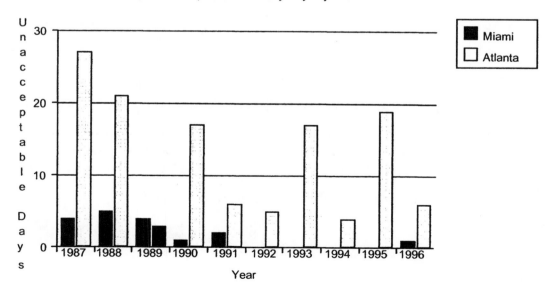

Which of the following correctly describes a condition depicted in the graph?
(Continued on next page)

A. In four years Atlanta had no days of unacceptable air quality.

B. In one year Miami had more unacceptable air quality days than Atlanta.

C. Miami's worst air quality year was better than Atlanta's best air quality year.

D. The trends show that air quality in Atlanta is improving each year.

10. The scatter plot below shows the amount of money earned (in millions of dollars) and the year of the concert tour for the twenty top-grossing North American concert tours over the years 1985 – 1997. *(Source: Pollstar, Fresno, CA.)*

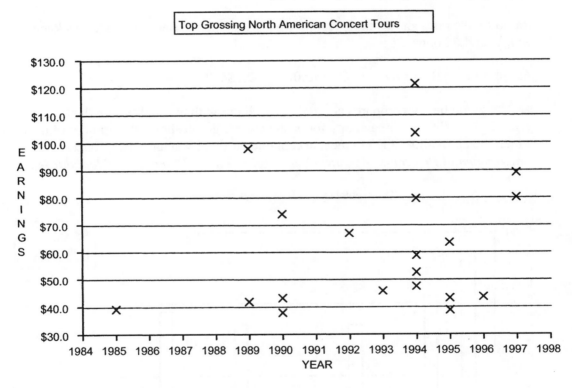

In which year did the third-highest grossing concert tour occur?

A. 1997

B. 1994

C. 1989

D. 1995

11. The scatter plot below relates the number of violent crimes reported to the number of property crimes reported by year over the years 1977 – 1996. *(Source: FBI Uniform Crime Reports 1996.)*

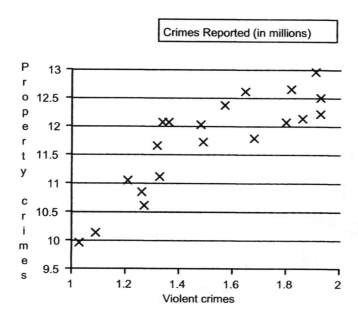

Approximately what is the greatest number of violent crimes reported in a year where the number of property crimes was less than 11.5 million?

A. 11.1 million

B. 1.35 million

C. 1.1 million

D. 1.9 million

CLAST SKILL I.D.1 *The student will identify information contained in bar, line and circle graphs*

See page 103 of this book to learn about this CLAST skill.

12. The bar graph below shows the revenues of the world's five largest corporations *(Source: Fortune Magazine).*

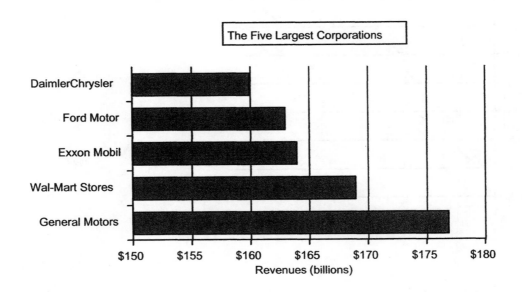

Approximately what is the difference in revenues between the fourth largest and the fifth largest corporations?

A. $3 billion

B. $300 billion

C. $17 billion

D. $13 billion

13. The line graph below shows the average temperature by month for Tampa, Florida.

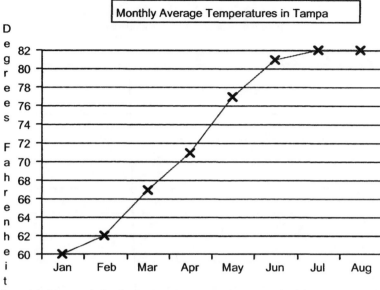

In which month is the average temperature roughly 77°?

A. April   B. May   C. June   D. March

14. The pie chart below summarizes the causes of home accident deaths in the U.S. in 1997 *(source: National Safety Council).*
Home Accident Deaths 1997 (in thousands)

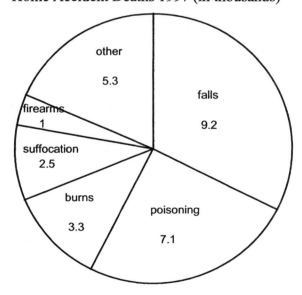

How many home accident deaths were caused by falls or burns?

A. 5,900   B. 12,500   C. 1,250   D. 16,300

15. The line graph below shows the number of new Broadway productions for several recent seasons *(source: League of American Theatres and Producers, Inc.).*

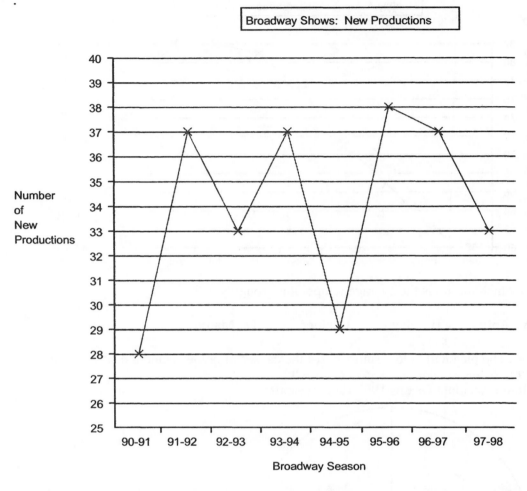

Broadway Shows: New Productions

In which season was the number of new productions ten greater than in the 90-91 season?

A. 91-92

B. 94-95

C. 97-98

D. 95-96

CLAST Exercises for Section 12.2 of *Thinking Mathematically.*

CLAST SKILL I.D.2 *The student will determine the mean, median and mode of a set of numbers.*

See page 106 of this book to learn about this CLAST skill.

1. What is the <u>median</u> of the data in the following sample?

12, 18, 12, 14, 12, 18, 12, 14, 15, 15

A. 14   B. 14.2   C. 12   D. 15

2. What is the <u>mean</u> of the data in the following sample?

4, 0, 3, 9, 3, 4, 3, 6

A. 3.5   B. 3   C. 4   D. 6

3. What is the <u>mode</u> of the data in the following sample?

30, 20, 20, 25, 20, 30, 25, 22

A. 22   B. 20   C. 25   D. 23.5

4. What is the <u>mean</u> of the data in the following sample?

10, 8, 12, 8, 10, 6, 16, 8, 8, 14, 8, 12

A. 8   B. 9   C. 10   D. 11

5. What is the <u>mode</u> of the data in the following sample?

0, 9, 0, 3, 1, 0, 1

A. 0   B. 1   C. 3   D. 2

6. What is the <u>median</u> of the data in the following sample?

4, 3, 5, 7, 12, 5, 4, 7, 7

A. 7   B. 12   C. 6   D. 5

7. What is the <u>mode</u> of the data in the following sample?

15, 15, 25, 15, 26, 25, 22, 24, 13

A. 20   B. 15   C. 22   D. 26

8. What is the <u>median</u> of the data in the following sample?

6, 1, 12, 3, 2, 10, 6, 0

A. 2.5   B. 6   C. 5   D. 4.5

9. What is the <u>mean</u> of the data in the following sample?

2, 15, 13, 4, 16, 7, 8, 14, 2

A. 16   B. 8   C. 2   D. 9

CLAST SKILL IV.D.1 *The student interprets real-world data involving frequency and cumulative frequency tables.*

See page 128 of this book to learn about this CLAST skill.

10. The table below contains information from the final round of the 1998 World Cup soccer championship, showing the distribution of games according to the total number of goals scored by both teams (goals scored on penalty kicks are not included).

Total number of goals scored	proportion
0	.0625
1	.1875
2	.0625
3	.4375
4	.0625
5	.1875

Find the median number of goals scored.

A. 2.5   B. .4375   C. 3   D. 2.8125

11. The table below shows the distribution according to age of the children in a pre-school program.

age	percent
3	60%
4	30%
5	10%

Find the mean age.

A. 4 years   B. 3 years   C. 33.3 years   D. 3.5 years

12. The table below shows the distribution of scores on a 10-point quiz.

quiz score	proportion
5	.10
6	.15
7	.30
8	.35
9	.05
10	.05

Find the median score.

A. 7.5   B. 7   C. 7.25   D. 8

13. A number of students were asked, "How many classes are you taking this semester?" The responses are summarized in the table below.

number of classes	percent of students
1	10%
2	10%
3	5%
4	35%
5	40%

Find the mean number of classes.

A. 3.85   B. 4   C. 3   D. 5.2

SECTION 12.2

14. For apartments in a student-oriented apartment complex the table below shows the distribution according to the number of residents per apartment.

number of residents	percent of apartments
1	18%
2	22%
3	21%
4	25%
5	14%

Find the mode for this distribution.

A. 4   B. 5   C. 3   D. 2.95

15. The table below shows the distribution of attendees at a dog show, according to the number of dogs owned.

number of dogs	proportion
0	.12
1	.36
2	.34
3	.16
4 or more	.02

Approximately what is the mean number of dogs owned?

A. 2.5   B. 1.72   C. 2   D. 1.6

16. The table below shows the distribution of households in a certain community, according to the number of bicycles owned per household.

number of bicycles	percent of households
0	22%
1	34%
2	35%
3 or more	9%

Find the median number of bicycles per household.

A. 2   B. 1.5

C. 1   D. 1.31

17. The table below shows the distribution of a community's rental housing units, according to the number of bathrooms per unit.

number of bathrooms	proportion
1	.25
1.5	.18
2	.22
2.5	.16
3	.14
3.5 or more	.05

Find the mode for this distribution.

A. .25   B. 3.5   C. 1   D. 2

CLAST SKILL II.D.1 *The student will recognize properties and interrelationships among the mean, median and mode in a variety of distributions.*

See page 111 of this book to learn about this CLAST skill.

18. A fraternal organization awards college scholarships in the amounts of $500, $1,000, $1,500 and $2,000. The distribution of awards is shown in the bar graph below.

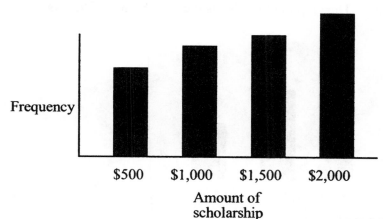

Which of the following statements is true about the distribution of awards?

A. The mean is the same as the median.

B. The median is greater than the mode.

C. The mean is greater than the mode.

D. The mode is greater than the median.

SECTION 12.2

19. The graph below shows the distribution of scores on a 10-point quiz.

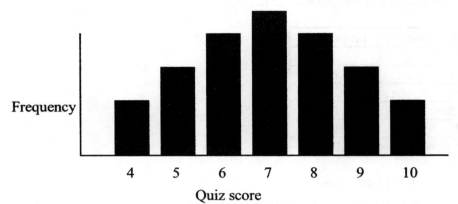

Which of the following statements is true about the distribution of scores?

A. The mean is less than the median.

B. The median is greater than the mode.

C. The mean is greater than the median.

D. The mode is the same as the mean.

20. Tickets for a minor league hockey game sell at the following price levels: $8.00, $11.50, $13.50 and $17.00. The graph below shows the distribution of tickets.

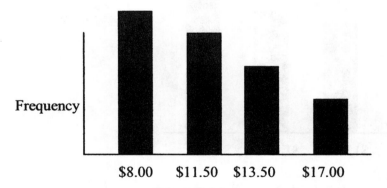

Which of the following statements is true about the distribution of ticket prices?

A. The mean is less than the mode.

B. The median is the same as the mean.

C. The mean is greater than the mode.

D. The mode is greater than the median.

21. A real estate developer is seeking permitting for a townhouse community. The community will feature three styles of luxury townhouse, ranging in price from $112,000 to $158,000. The graph below shows the planned distribution of townhouses.

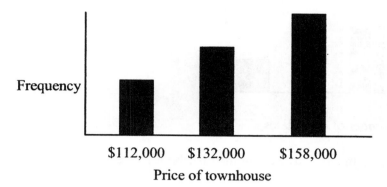

Which of the following statements is true about the distribution of townhouse prices?

A. The median is greater than the mode.

B. The mean is the same as the mode.

C. The mode is greater than the median.

D. The mode is less than the mean.

22. The graph below shows the distribution of a number of middle school students according to the number of "A" grades on their report cards.

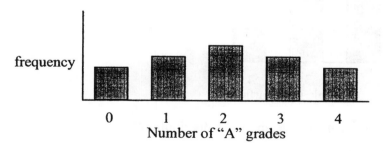

Which of the following statements is true about the distribution of "A" grades?

A. The median is less than the mean.

B. The median is less than the mode.

C. The mean is the same as the median.

D. The mode is greater than the mean.

SECTION 12.2

23. The graph below shows the distribution of primary election contests according to the number of candidates in each contest, for a certain community.

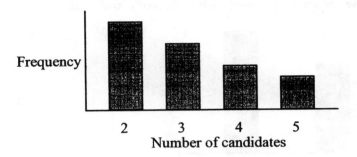

Which of the following statements is true about the distribution of election contests?

A. The mean is less than the mode.

B. The mode is the same as the mean.

C. The mean is greater than the median

D. The median is less than the mode.

24. The graph below shows the distribution of apartments according to the number of people living in each apartment at a certain apartment complex.

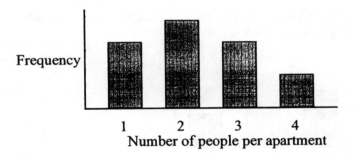

Which of the following statements is true about the distribution?

A. The mean is greater than the mode.

B. The median is less than the mode.

C. The mean is the same as the mode.

D. The mean is less than the mode.

25. When he did a belly flop while attempting a dive during try-outs for the U.S. Olympic dive team, half of the judges gave Omar a score of 3. Of the others, most gave a score of 2 while a few gave a score of 1. Which of the following statements is true about the distribution of scores?

A. The median is less than the mean.

B. The mean is greater than the mode.

C. The median is greater than the mode.

D. The mean is less than the mode.

26. Half of the households in Tom's neighborhood own two cars. The others are evenly divided between those that own one car and those that own three cars. Which of the following statements is true about the distribution of households according to number of cars owned?

A. The mean is the same as the median.

B. The median is less than the mode.

C. The mean is less than the mode.

D. The median is greater than the mean.

27. An insurance agent finds that almost half of the applicants for auto insurance report having no tickets for moving violations during the past three years. Of the others, a lesser number report having had one such moving violation, fewer still report two moving violations and fewer still report having three or more moving violations. Which of the following statements is true about this distribution?

A. The mean is less than the mode.

B. The mode is the same as the median.

C. The median is the same as the mean.

D. The mode is less than the mean.

CLAST Exercises for Section 12.4 of *Thinking Mathematically.*

CLAST SKILL IV.D.1 *The student interprets real-world data involving frequency and cumulative frequency tables.*

See page 128 of this book to learn about this CLAST skill.

1. The table below shows the percentile distribution of students according to their scores on a math exam.

Test Score	Percentile Rank
100	98
95	92
90	85
80	70
70	55
65	50
55	35
45	20

What percentage of students scored less than 55?

A. 55%   B. 35%   C. 20%   D. 80%

2. The table below shows the percentile distribution according to length of spotted sea trout harvested by recreational anglers in a Florida Gulf Coast fishery.

Length of fish (inches)	Percentile Rank
28	99
25	95
23	80
21	75
19	50
18	35
16	15

What percentage of these fish are 25 inches long or longer?

A. 5%   B. 99%

C. 1%   D. 95%

3. The table below gives the percentile distribution according to value of single-family homes in a certain community.

Value of Home ($)	Percentile Rank
250,000	95
150,000	80
125,000	65
110,000	50
100,000	45
85,000	35
60,000	25
50,000	10

What percentage of homes have values between $50,000 and $100,000?

A. 35%

B. 60%

C. 70%

D. 25%

4. The table below shows the percentile distribution according to average monthly utility bill of the residential customers of a municipal utility company.

Average Monthly Bill ($)	Percentile Rank
$350	95
$250	75
$175	50
$150	30
$100	25

What percentage of customers have average monthly utility bills of $250 or more?

A. 75%

B. 95%

C. 170%

D. 25%

SECTION 12.4

5. The table below shows the projected percentile distribution according to age of the U.S. population in the year 2010 *(source: Bureau of the Census, U.S. Department of Commerce).*

Age (years)	Percentile Rank
85	98
65	87
55	75
45	61
35	48
25	35
18	24
5	7

According to these projections, what percentage of the U.S. population will be younger than 25 years of age in the year 2010?

A. 24%   B. 31%

C. 35%   D. 66%

6. The table below shows the percentile distribution according to age of licensed drivers in the U.S. *(source: U.S. Department of Transportation).*

Age of Driver (years)	Percentile Rank
75	98
65	88
55	77
45	60
35	37
25	14
20	5

What percentage of drivers are between 20 and 45 years of age?

A. 55%   B. 51%

C. 60%   D. 37%

CLAST Exercises for Section 12.5 of *Thinking Mathematically.*

CLAST SKILL III.D.1 *The student infers relations and makes accurate predictions from studying statistical data.*

See page 124 of this book to learn about this CLAST skill.

1. The scatter plot below relates public high school graduation rates to fourth grade performance on a reading test for a number of states.

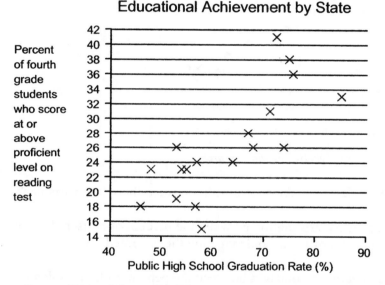

Educational Achievement by State

Source: National Center For Education Statistics, U.S. Dept. of Education.

Which of the following best describes the relationship between fourth grade reading proficiency and high school graduation rates?

A. Higher fourth grade reading proficiency tends to be associated with lower high school graduation rates.

B. There is no apparent association between high school graduation rates and fourth grade reading proficiency.

C. Higher high school graduation rates tend to be associated with higher fourth grade reading proficiency.

D. Higher fourth grade reading proficiency causes higher high school graduation rates.

2. The graph below relates the percentage of U.S. households having cable TV service to the rating (percent of households viewing) of the year's top-rated TV show for a number of recent years.

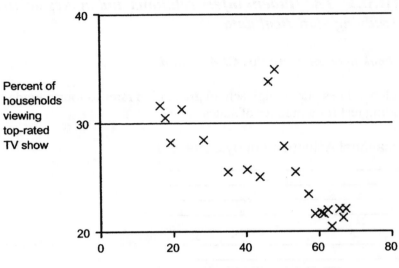

Source: *Nielsen Media Research*

Which of the following best describes the relationship between these two variables?

A. There appears to be a positive association between the percent of households with cable TV service and the percent of households viewing the top-rated TV show.

B. There appears to be a negative association between the percent of households with cable TV service and the percent of households viewing the top-rated TV show.

C. There is no apparent association between the percent of households with cable TV service and the percent of households viewing the top-rated TV show.

D. Increasing rates of cable TV subscribership tend to be associated with increasing ratings for the top-rated TV show.

3. The scatter plot below relates the enrollment to the typical annual resident tuition for a number of institutions of Florida's State University System.

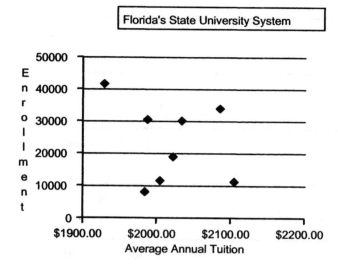

Which of the following best describes the relationship between these two variables?

A. Higher enrollments tend to be associated with higher tuition.

B. Higher enrollments cause higher tuition.

C. There is a negative association between enrollment and tuition.

D. There is no apparent association between enrollment and tuition.

4. The scatter plot below shows the relationship between annual sales of music CDs and music cassettes for a number of recent years.

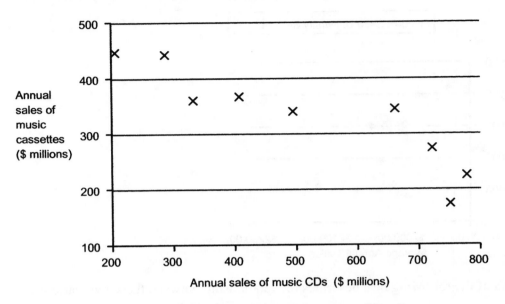

Source: Recording Industry Association of America

Which of the following best describes the relationship between these two variables?

A. Increasing CD sales are associated with decreasing cassette sales.

B. Increasing CD sales are associated with increasing cassette sales.

C. There is no apparent association between CD sales and cassette sales.

D. There is a positive correlation between CD sales and cassette sales.

5. The scatter plot below shows the relationship between average SAT math scores and average SAT verbal scores for a number of states.

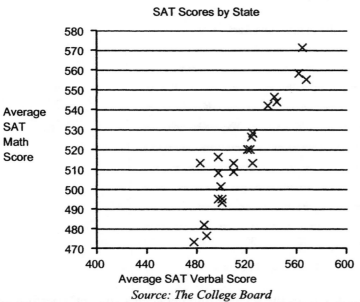

SAT Scores by State

Source: *The College Board*

Which of the following best describes the relationship between SAT math and verbal scores?

A. There is a positive association between math and verbal scores.

B. There is a negative association between math and verbal scores.

C. High math scores tend to cause high verbal scores.

D. People with good verbal skills tend to have poor math skills.

SECTION 12.5

6. The scatter plot below shows, for the students in one section of Liberal Arts Math, the relationship between the student's overall score in the course and the number of semester hours of coursework the student is taking.

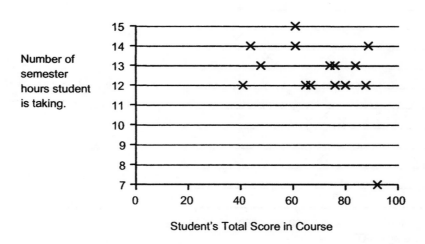

Which of the following best describes the relationship between a student's performance in the course and the number of semester hours the student is taking?

A. Students who are taking more semester hours tend to get higher scores in the course.

B. Students who are taking more semester hours tend to get lower scores in the course.

C. There is no apparent association between a student's score in the course and the number of semester hours the student is taking.

D. Good students take more classes than poor students do.

Practice CLAST A

1. Identify the missing term from the following geometric progression.

−1, 1/3, −1/9, 1/27, −1/81, ____

A. 1/243   B. −1/243   C. −1/3   D. 1/108

2. Jeremy has purchased five textbooks. The least expensive book cost $24 and the most expensive book cost $70. Which of the following could be a reasonable estimate of the total cost of the five books?

A. $120   B. $85   C. $200   D. $350

3. The circle graph below shows preferences of a number of fast food consumers. What percent prefer something other than chicken?

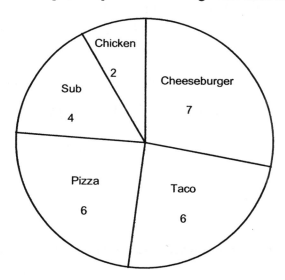

A. 2%

B. 98%

C. 8%

D. 92%

4. Referring to the figure below, select the statement that is true. (The measure of angle ABC is represented by "x.")

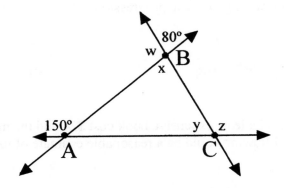

A. z = 150°   B. w = 80°   C. y = 70°   D. x = 30°

5. Select the figure that must possess all of the following characteristics.

i. is a quadrilateral;
ii. all sides have equal length;
iii. diagonals have equal length.

A. square   B. rhombus   C. trapezoid   D. rectangle

6. Find $f(3)$, given $f(x) = 4x - 2x^2$

A. 6   B. –6   C. 30   D. 3

7. How many positive integers leave a remainder of 3 when divided into 31 and leave a remainder of 8 when divided into 50?

A. 0   B. 1   C. 2   D. 14

8. Select the statement that is the negation of the statement "If you like pasta, then you like linguine."

A. If you like linguine, then you like pasta.
B. If you don't like pasta then you don't like linguine.
C. You like pasta and you don't like linguine.
D. If you like pasta, then you don't like linguine.

9. Round the measure 6.4548 centimeters to the nearest tenth of a centimeter.

A. 6.0 cm   B. 6.5 cm   C. 60 cm   D. 6.4 cm

10. Select the correct expanded notation for 600.000 07.

A. $\left(6 \times 10^3\right) + \left(7 \times \dfrac{1}{10^5}\right)$ B. $\left(6 \times 10^2\right) + \left(7 \times \dfrac{1}{10^4}\right)$

C. $\left(6 \times 10^3\right) + \left(7 \times \dfrac{1}{10^4}\right)$ D. $\left(6 \times 10^2\right) + \left(7 \times \dfrac{1}{10^5}\right)$

11. Choose the correct solution set for the system of linear equations.
$8x - 3y = -4$
$12x + y = 5$

A. $\{(1, 4)\}$ B. $\left\{\left(\dfrac{1}{4}, 2\right)\right\}$

C. $\{(1, -7)\}$ D. $\{(x, y)| y = -12x + 5\}$

12. Six tennis players are going to play a "round-robin" tournament, meaning that each player will play one match against each of the other players. How many matches will be held?

A. 30   B. 36   C. 15   D. 12

13. The formula $C = 2x + 25$ gives the total cost (C, in $) of manufacturing x pounds of chocolate syrup. How many pounds of syrup can be manufactured for $125?

A. 150   B. 250   C. 50   D. 275

14. Select the rule of logical equivalence which directly (in one step) transforms statement "i" into statement "ii."

i. If x is an odd number, then $3x$ is an odd number.
ii. x is not an odd number or $3x$ is an odd number.

A. "Not (p or q) is equivalent to "not p and not q."
B. "If p, then q" is equivalent to "If not q, then not p."
C. "If p, then q" is equivalent to "If not p, then not q."
D. Correct equivalence is not given.

15. $\left(\dfrac{3}{5}\right)^4 =$

A. $\left(\dfrac{4 \times 4 \times 4}{4 \times 4 \times 4 \times 4 \times 4}\right)$

B. $\left(\dfrac{3}{5}\right) \times \left(\dfrac{3}{5}\right) \times \left(\dfrac{3}{5}\right) \times \left(\dfrac{3}{5}\right)$

C. $\dfrac{3}{5} + \dfrac{3}{5} + \dfrac{3}{5} + \dfrac{3}{5}$

D. $\dfrac{3 \times 4}{5 \times 4}$

16. Identify the conditions corresponding to the shaded region of the plane.

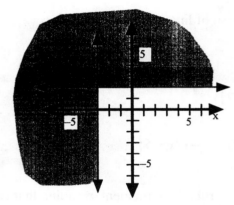

A. $x \le -3$ and $y \ge 2$

B. $y \le -3$ and $x \ge 2$

C. $y \le -3$ or $x \ge 2$

D. $x \le -3$ or $y \ge 2$

17. All of the arguments A - D have true conclusions, but one of the arguments is not valid. Select the argument that is **not** valid.

A. If spinach is a green leafy vegetable then it is a source of dietary iron. Spinach is a green leafy vegetable. Therefore spinach is a source of dietary iron.

B. If grapefruit is a citrus fruit, then it is a source of vitamin C. Grapefruit is a source of vitamin C. Therefore, grapefruit is a citrus fruit.

C. If orange juice is a seafood, then it is a source of protein. Orange juice is not a source of protein. Therefore, orange juice is not a seafood.

D. If beans are a source of dietary fiber, then they are good for your health. Beans are a source of dietary fiber. Therefore, beans are good for your health.

18. $2\dfrac{1}{4} \div \dfrac{1}{3} =$ A. $\dfrac{1}{6}$ B. $\dfrac{3}{4}$ C. $6\dfrac{3}{4}$ D. 6

19. Find the correct solutions to

$2x^2 - x - 2 = 0$

A. $\dfrac{1+\sqrt{17}}{4}$ and $\dfrac{1-\sqrt{17}}{4}$

B. $\dfrac{1+\sqrt{15}}{2}$ and $\dfrac{1-\sqrt{15}}{2}$

C. $\dfrac{-1+\sqrt{17}}{2}$ and $\dfrac{-1-\sqrt{17}}{2}$

D. $\dfrac{-1+\sqrt{17}}{4}$ and $\dfrac{-1-\sqrt{17}}{4}$

20. Study the information given below. If a logical conclusion is given, select that conclusion. If none of the conclusions given is warranted, select the option expressing this condition.

All alligators are dangerous. No kittens are dangerous. Fluffy is not a kitten.

A. Fluffy is an alligator.

B. Fluffy is not dangerous.

C. Fluffy is not an alligator.

D. None of the above is warranted.

21. What is the area of a circular region with a diameter of 10 meters?

A. 100π square meters

B. 10π square meters

C. 5π square meters

D. 25π square meters

22. At Sudsy Stanley's Car Wash it costs $3.95 to wash a car and another $5.95 to wax a car that has been washed. Fredo has four taxi cabs that will be brought to the car wash. All four will be washed and two will be waxed. Find the total cost.

A. $21.75   B. $19.80   C. $27.70   D. $31.70

23. Half of the students who took a five-point quiz received scores of 3. Of the others, most received scores of 4 while a few received scores of 5. Which of the following statements is true about the distribution of scores?

A. The mean is greater than the mode.

B. The median is the same as the mean.

C. The median is the same as the mode.

D. The mean is the same as the mode.

24. The table below shows the distribution of students according to the number of minutes required for them to complete their final exam.

Number of minutes	Percent
0–30	25%
31 – 60	45%
61 – 90	20%
91 – 120	8%
more than 120	2%

If two students are randomly selected, find the probability that they both finished the exam in 30 minutes or less.

A. .50   B. .25   C. .4375   D. .0625

25. Which is a linear factor of the following expression?

$5x^2 - 19x - 4$

A. $x - 4$ B. $5x - 4$ C. $5x + 4$ D. $x + 1$

26. If $2x + 1 > 6x + 5$, then

A. $x < 1$ B. $x > -1$ C. $x < -1$ D. $x > 1$

27. Select the statement below that is logically equivalent to "If you are in Miami then you are in Dade County."

A. You aren't in Miami and you aren't in Dade County.

B. If you aren't in Miami then you aren't in Dade County.

C. If you are in Dad County then you are in Miami.

D. If you aren't in Dade County then you aren't in Miami.

28. Find the <u>median</u> of the data in the following sample.

18, 10, 6, 6, 18, 6,10, 6

A. 6   B. 10   C. 8   D. 12

29. A rectangular flowerbed that is 3 yards wide, 10 yards long and 9 inches deep will be filled with topsoil that costs $12 per cubic yard. Find the cost of the topsoil.

A. $90   B. $3,240   C. $1,080   D. $270

30. $\sqrt{6} \times \sqrt{15} =$

A. $\sqrt{21}$ B. 6×15 C. $16\sqrt{5}$ D. $4\sqrt{5}$

31. $(-.03) \times (-.6) =$

A. .18   B. −.18   C. .018   D. −.018

32. If 50 is decreased to 40, what is the percent decrease?

A. 20%   B. 16%   C. 10%   D. 25%

33. $\dfrac{1}{4} - \dfrac{3}{4} \div \dfrac{1}{2} \times \dfrac{1}{4} =$

A. $\dfrac{7}{32}$ B. $-5\dfrac{1}{4}$ C. $-\dfrac{1}{4}$ D. $-\dfrac{1}{8}$

34. Sets A, B, C and U are related as shown in the diagram. Which of the following is true, assuming that no region of the diagram is empty?

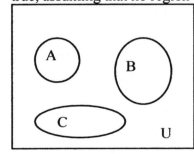

A. Any element that is not a member of A is a member of B or C.

B. There is no element that is a member of A and B.

C. Any element that is a member of U is not a member of A or B or C.

D. None of the above is true.

35. 175% =   A. 175/1000   B. 17.5   C. 1.75   D. .00175

36. Identify the symbol that should be placed in the blank to form a true statement.

   A. <   B. >   C. =

37. Select the figure in which all of the triangles are similar.

A.

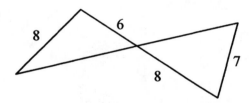

B.

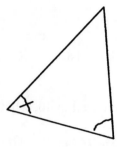

C.

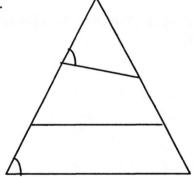

D.

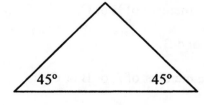

38. For each of the statements below, determine whether $x = 3$ is a solution.

i. $|x| < 3$

ii. $2x^2 - 7x = -3$

iii. $x + 3 = x - 3$

A. i only   B. i and ii only   C. ii only   D. i, ii and iii

39. The formula for finding the simple interest (I) on a loan is $I = PRT$. How much interest will Bill pay on his property purchase if he finances \$12,000 (P) at a 10% simple interest rate $(R = .10)$ for 3 years (T)?

A. \$1,200   B. \$4,000   C. \$3,600   D. \$15,600

40. Study the information in the figures below. In each figure, A represents the area of the shaded part of the figure.

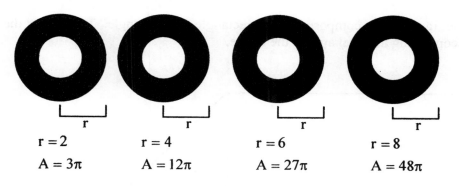

$r = 2$ $r = 4$ $r = 6$ $r = 8$

$A = 3\pi$ $A = 12\pi$ $A = 27\pi$ $A = 48\pi$

Find the area (A) of a similar figure if $r = 10$.

A. 75π B. 15π C. 51π D. 49π

41. The square of a number, decreased by 3 times the number, is equal to the number. Which equation could be used to find x, the number?

A. $-3x^2 = x$ B. $x^2 - 3x = x$ C. $x^2 - 3 = x$ D. $x^2 - 3 = x^2$

42. Select the conclusion that will make the following argument valid.

If all of us are happy we will have a party. We won't have a party.

A. None of us is happy.

B. If we have a party then all of us will be happy.

C. To heck with those who aren't happy, the rest of us will still have a party.

D. Some of us aren't happy.

43. A manufacturer produces 120-minute blank videocassette tapes. Due to random errors in the manufacturing process ten percent of the videocassette tapes contain less than 115 minutes of tape. If two videocassettes are randomly selected find the probability that at least one of them contains less than 115 minutes of tape?

A. .19   B. .2   C. .01   D. .1

44. Study the figure below composed from a rectangle and a triangle. Then select the formula for computing the total area of the figure.

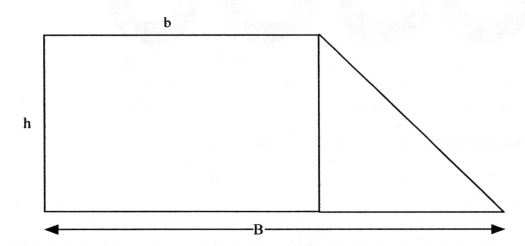

A. Area = hb + (1/2)hB

B. Area = hb + (1/2)(B – b)h

C. Area = hB + hb

D. Area = (1/2)(hBb)

45. Select the units of measure that would be appropriate for measuring the height of the cylinder shown.

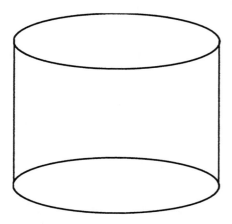

A. square inches   B. millimeters   C. cubic centimeters   D. square feet

46. Moe and Joe are both going to drive from Heretown to Thereville (see figure below). Joe is going to travel the direct 10-mile road. Moe would rather travel the alternative scenic route, however, traveling westward to Middleton and then northward to Thereville. If Moe can average 56 miles per hour, how long will it take him to make the journey?

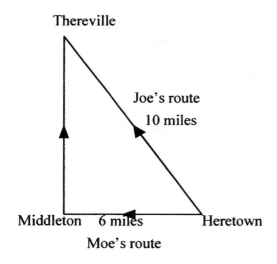

A. 7 hours

B. 1/4 hour

C. 1/7 hour

D. 1/12 hour

47. The relative frequency table below shows the distribution according to height of the students in a gym class.

height (inches)	Percent of students
43	4%
44	8%
45	18%
46	16%
47	15%
48	22%
49	14%
50 or more	3%

Approximately what is the mean height?

A. 47 inches   B. 46 inches   C. 46.67 inches   D. 46.5 inches

48. The student government wants to conduct a survey of students to determine the extent of support for spending student activity fees to fund intramural sports. Which of the following would be most appropriate for selecting an unbiased sample?

A. Distribute survey forms to 100 people attending an intramural softball game.

B. Place survey forms on 100 bulletin boards around campus.

C. Mail survey forms to 100 people randomly selected from an alphabetical list of all students.

D. Give survey forms to 100 people randomly selected from the local telephone directory.

49. $9,000,000 \times 0.00003 =$

A. .27

B. 27

C. 2.7

D. 270

50. Given that:

i. If you are a cat, then you eat mice; and

ii. You aren't a cat;

determine which conclusion can be logically deduced.

A. You are a cat.   B. You eat mice.

C. You don't eat mice.   D. None of the above.

51. 96 is 40% of what number?

A. 1.25   B. 38.4   C. 240   D. 126

52. Choose the expression equivalent to the following.

6a + 4b

A. 6b + 4a   B. 4b + 6a   C. 10(a + b)   D. 10ab

53. The scatter plot below shows the relationship between average number of students per teacher in public schools and average annual salary ((in $ thousands) for public school teachers for a number of southern states.

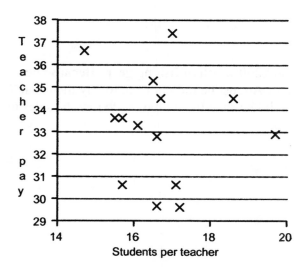

Which of the following best describes the relationship between teacher salary and number of students per teacher? (Continued on next page.)

A. There is a positive correlation between teacher salary and number of students per teacher.

B. There is a negative correlation between teacher salary and number of students per teacher.

C. Having higher salaries causes teachers to have more students.

D. There is no apparent association between teacher salary and number of students per teacher.

54. Three workers can process 500 pounds of seafood in 4 hours. Let x be the number of pounds of seafood that they can process in 6 hours. Select the correct statement of the given condition.

A. $\dfrac{x}{3} = \dfrac{500}{4}$

B. $\dfrac{x}{4} = \dfrac{3}{500}$

C. $\dfrac{x}{4} = \dfrac{6}{500}$

D. $\dfrac{x}{6} = \dfrac{500}{4}$

55. Jerry bought $80 worth of clothes and $200 worth of fishing gear. Because he bought these things during the first week of August he did not have to pay the usual 7% state sales tax on the clothes. How much did he save by not having to pay the sales tax?

A. $5.60

B. $14.00

C. $19.60

D. $85.60

Practice CLAST B

1. 60 is what percent of 40?

A. 120%   B. $66\frac{2}{3}$%   C. 1.5%   D. 150%

2. Select the expression equivalent to the following: $3a + 9b$

A. $3(a + 3b)$ B. $3(a + 9b)$ C. $12(a + b)$ D. $12ab$

3. The scatter plot below relates the average composite ACT score to the percent of high school graduates taking the ACT for a number of states. *(Source: ACT, Inc)*

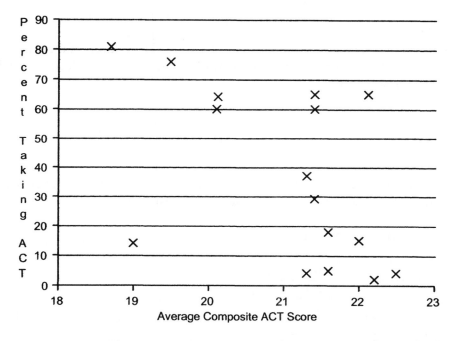

Which of the following best describes the relationship between ACT scores and percent of students taking the ACT?

A. Higher percentages of students taking the ACT tend to be associated with lower scores.

B. Higher percentages of students taking the ACT tend to be associated with higher scores.

C. Higher percentages of students taking the ACT cause lower scores.

D. Higher percentages of students taking the ACT cause higher scores.

357

PRACTICE CLAST B

4. Six rottweilers can consume 50 pounds of dog chow in 10 days. Let x be the number of pounds of dog chow that 8 rottweilers can consume in 10 days. Select the correct statement of the condition.

A. $\dfrac{6}{x} = \dfrac{50}{10}$ 　　 B. $\dfrac{6}{x} = \dfrac{10}{50}$ 　　 C. $\dfrac{6}{x} = \dfrac{50}{8}$ 　　 D. $\dfrac{6}{50} = \dfrac{8}{x}$

5. Ed weighs 180 pounds and Ned weighs 25% more than Ed. How much does Ned weigh?

A. 45 pounds 　　 B. 225 pounds 　　 C. 205 pounds 　　 D. 210 pounds

6. A thirteen-foot extension ladder is leaning against a wall. The bottom of the ladder is resting at a place that is 5 feet from the base of the wall. How high is the place where the top of the ladder touches the wall?

A. 10 feet 　　 B. 18 feet 　　 C. 12 feet 　　 D. 8 feet

7. The table below shows the percentile distribution according to length of spotted sea trout harvested by recreational anglers in a Florida Gulf Coast fishery.

Length of fish (inches)	Percentile Rank
28	99
25	95
23	80
21	75
19	50
18	35
16	15

What percentage of these fish are less than 18 inches long?

A. 15%

B. 65%

C. 50%

D. 35%

8. A national Internet service provider wants to conduct a telephone survey of its subscribers to determine the extent to which subscribers are satisfied with the service. Which method would be most appropriate for selecting an unbiased sample?

A. Randomly select 5000 names from an alphabetical list of all subscribers.

B. Randomly select 500 names from a national list of certified web technicians.

C. Select the first 500 names from a national list of all subscribers.

D. Randomly select 500 names from a list of subscribers who have filed complaints or had late payments.

9. $420 \div .000000006 =$

A. 7×10^8 B. 7×10^{-8} C. 7×10^{10} D. 7×10^{-10}

10. Given that:

i. If you forget your pom-poms then you can't go to cheerleading class; and

ii. You can't go to cheerleading class;

determine which conclusion can be logically deduced.

A. You forgot your pom-poms.   B. You didn't forget your pom-poms.

C. You overslept.   D. None of the above.

11. Look for a common linear relationship between the numbers in each pair. Then identify the missing term.

$(4, 1)$ $(2.4, .6)$ $(-12, -3)$ $(40, 10)$, $(4/3, \underline{\quad})$

A. $-1/3$ B. $1/3$ C. $1/12$ D. $-1/12$

12. Two hundred patients were weighed before visiting with their physicians. All of the male patients weighed more than 120 pounds but less than 300 pounds. All of the female patients weighed more than 85 pounds but less than 250 pounds. Which of the following could be a reasonable estimate of the average weight of a male patient?

A. 116 pounds   B. 320 pounds   C. 190 pounds   D. 100 pounds

PRACTICE CLAST B

13. The bar graph below shows the distribution according to age of the children in a summer camp. For what age is it true that 10 campers are more than that age?

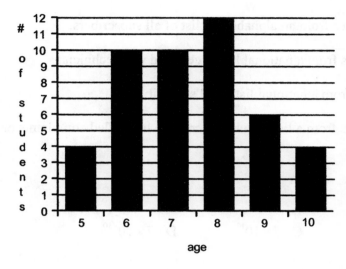

age

A. 8   B. 9   C. 6   D. 7

14. In the figure below $\overline{AB} \| \overline{DC}$. Select the statement that is true.

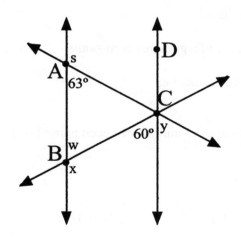

A. △ABC is equilateral.   B. w = y

C. s = x   D. x = 120°

15. Referring to the figure below, select the pair of complementary angles.

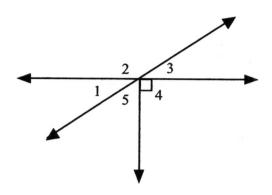

A. 2 and 4   B. 2 and 3   C. 5 and 3   D. 3 and 4

16. Find $f(-2)$, given $f(x) = 5 + x - x^2$

A. 1   B. 2   C. 7   D. -1

17. Find the greatest factor of 108 that is also a divisor of 72.

A. 216   B. 27   C. 12   D. 36

18. Select the statement that is the negation of the statement "Some baseball players are shortstops."

A. All baseball players are shortstops.
B. Some baseball players aren't shortstops.
C. No baseball players are shortstops.
D. All shortstops are baseball players.

19. Round the diameter of the washer to the nearest 1/2-inch.

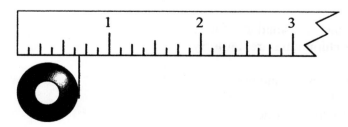

A. 1 inch   B. 1/2 inch   C. 3/2 inch   D. 0 inch

PRACTICE CLAST B

20. Select the place value associated with the underlined digit.

3,955,0<u>6</u>9,300.007 455 2

A. 10^4 B. 10^5 C. $\dfrac{1}{10^5}$ D. 10^6

21. Choose the correct solution set for the system of linear equations.
$2x - 6y = 6$
$x - 3y = 1$

A. $\{(0, -1)\}$ B. $\{(4, 1)\}$

C. $\left\{ (x, y) \Big| y = -\dfrac{1}{3}(x - 1) \right\}$ D. The empty set

22. There are four candidates for the job of napkin folder. A committee will examine their credentials and interview each of the candidates. Afterward, the committee will rank the four candidates from "best suited" to "worst suited" for the job. How many outcomes are possible in this ranking process?

A. 24   B. 16   C. 8   D. 12

23. Ervin is unhappy because his truck only gets 8 miles per gallon of gasoline, so he has purchased a gasoline additive that is supposed to improve gas mileage. The instructions say to use three ounces of additive for every five gallons of gasoline. If he pours 10 ounces of additive into his gas tank, how many gallons of gasoline should he include?

A. 6 gallons   B. 15/8 gallons   C. $16\dfrac{2}{3}$ gallons   D. $13\dfrac{1}{3}$ gallons

24. Select the rule of logical equivalence which directly (in one step) transforms statement "i" into statement "ii."

i. It is not the case that chipmunks are small and furry.
ii. Chipmunks are not small or chipmunks are not furry.

A. "Not (p or q)" is equivalent to "not p and not q."
B. "Not (p and q)" is equivalent to "not p or not q."
C. "If p, then q" is equivalent to "not p or not q."
D. Correct equivalence is not given.

25. $2^5 \times 3^2 =$

A. $(5)^{10}$

B. $(5 \cdot 5) \times (2 \cdot 2 \cdot 2)$

C. $(2 \cdot 2 \cdot 2 \cdot 2 \cdot 2) \times (3 \cdot 3)$

D. $(6 \cdot 6 \cdot 6 \cdot 6 \cdot 6 \cdot 6)$

26. Select the graph whose shaded region corresponds to the conditions $x + 2y < 6$.

A.

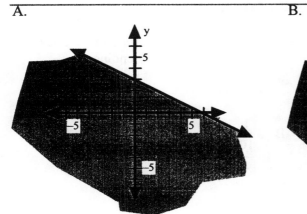

B.

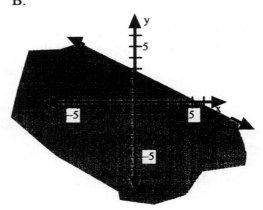

C.

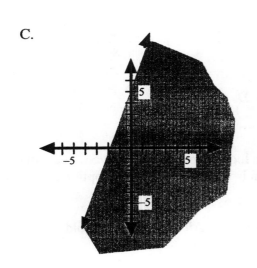

D.

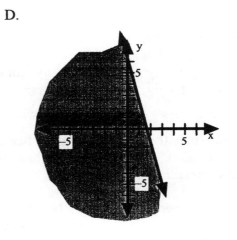

27. All of the arguments A - D have true conclusions, but one of the arguments is not valid. Select the argument that is **not** valid.

A. All of Shakespeare's comedies have happy endings and *Much Ado About Nothing* is one of Shakespeare's comedies. Therefore, *Much Ado About Nothing* has a happy ending.

B. All of Shakespeare's tragedies have unhappy endings and *As You Like It* doesn't have an unhappy ending. Therefore, *As You Like It* is not one of Shakespeare's tragedies.

C. All of Shakespeare's comedies have happy endings and *Henry V* has a happy ending. Therefore, *Henry V* is one of Shakespeare's comedies.

D. All of Shakespeare's tragedies have unhappy endings and *MacBeth* is one of Shakespeare's tragedies. Therefore, *MacBeth* has an unhappy ending.

28. $1\frac{1}{2} + 2\frac{3}{5} =$ A. $3\frac{4}{7}$ B. $3\frac{1}{10}$ C. $4\frac{1}{10}$ D. $3\frac{3}{7}$

29. Find the correct solutions to

$$8x^2 - 6x + 1 = 0$$

A. $\frac{1}{2}$ and $\frac{1}{4}$

B. $\frac{3+\sqrt{17}}{8}$ and $\frac{3-\sqrt{17}}{8}$

C. $\frac{3+\sqrt{17}}{4}$ and $\frac{3-\sqrt{17}}{4}$

D. 1 and $\frac{1}{2}$

30. Study the information given below. If a logical conclusion is given, select that conclusion. If none of the conclusions given is warranted, select the option expressing this condition.

All bold people are heroic. No cowards are heroic. All astronauts are bold.

A. All heroic people are astronauts.

B. No heroic people are astronauts.

C. No astronauts are cowards.

D. None of the above is warranted.

31. Find the volume of a rectangular solid that is 2 yards long, 1 yard wide and 1 yard high.

A. 2/3 feet   B. 54 feet   C. 54 cubic feet   D. 2/3 cubic feet

32. Jack's employer offers a group life insurance policy for employees and their spouses. For a person whose age is in the 19 to 37 year range, the monthly rate for each policy is 15¢ for each $1,000 worth of insurance purchased; for people in the 38 to 59 year range the monthly rate for each policy is 25¢ for each $1,000 worth of insurance purchased. Jack is going to purchase a $100,000 policy for himself and a $100,000 policy for his wife. Jack is 39 years old and his wife is 36 years. What will be the total monthly rate for the two policies?

A. $40   B. $400   C. $50   D. $20

33. At a certain community college on the first day of classes all of the sections of MGF1107 have enrollments ranging from 26 to 30 students. The graph below shows the distribution of sections according to enrollment.

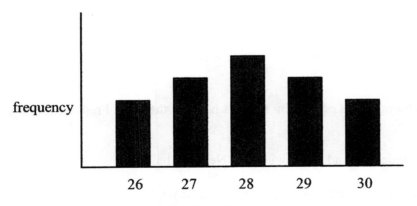

Number of students per section

Which of the following statements is true about this distribution?

A. The median is less than the mode.

B. The mean is the same as the median.

C. The mode is less than the mean.

D. The mode is greater than the mean.

34. The table below shows the distribution of students according to the number of minutes required for them to complete their final exam.

Number of minutes	Percent
0–30	25%
31 – 60	45%
61 – 90	20%
91 – 120	8%
more than 120	2%

If one student is randomly selected, find the probability that he or she finished the test in 31 minutes or more, given that he or she finished the test in 90 minutes or less.

A. 65/100   B. 65/75   C. 65/90   D. 45/65

35. Which is a linear factor of the following expression? $6x^2 - x - 5$

A. $x + 5$ B. $6x - 1$ C. $6x + 5$ D. $x + 1$

36. If $3x + 1 = 2x + 7$, then

A. $x = 8/5$ B. $x = 6$ C. $x = 6/5$ D. $x = 8$

37. Select the statement that is equivalent to "It is not true that both Fred and Barney are Neanderthals."

A. Fred is not a Neanderthal and Barney is not a Neanderthal.

B. Fred is not a Neanderthal or Barney is not a Neanderthal.

C. Fred is a Neanderthal or Barney is a Neanderthal.

D. Fred is a Neanderthal and Barney is not a Neanderthal.

38. Find the <u>mean</u> of the data in the following sample.

24, 8, 20, 24, 10, 6, 16, 8, 24, 24

A. 8   B. 24   C. 16.4   D. 18

39. How many 3-foot by 3-foot square concrete slabs are needed to pave a rectangular walkway that is 2 yards wide and 27 yards long?

A. 54   B. 27   C. 6   D. 18

40. $\dfrac{7}{\sqrt{2}} =$ A. $\dfrac{\sqrt{7}}{2}$ B. $\dfrac{7\sqrt{2}}{4}$ C. $\dfrac{7\sqrt{2}}{2}$ D. $7\sqrt{2}$

41. $2.61 + .439 =$ A. .3049   B. 3.049   C. 3.49   D. .349

42. If you increase 6 by 25% of itself, what is the result?

A. 7.5   B. 1.5   C. 31   D. 24

43. $\dfrac{1}{8} + \dfrac{1}{2} \times \dfrac{2}{3} =$ A. $\dfrac{5}{12}$ B. $\dfrac{2}{11}$ C. $\dfrac{11}{24}$ D. $\dfrac{1}{11}$

44. Sets A, B, C and U are related as shown in the following diagram.

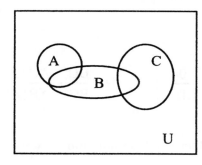

Select the statement that correctly describes a relationship between the sets in the diagram, assuming that no region of the diagram is empty.

A. Any element that is a member of U is a member of A or B or C.

B. Any element that is a member of U is not a member of A or B or C.

C. Any element that is a member of A and B is not a member of C.

D. None of the above is true.

45. $.088 =$ A. $\dfrac{88}{1000}\%$ B. 8.8% C. $\dfrac{88}{100}$ D. 88%

46. Identify the symbol that should be placed in the blank in order to form a true statement.

$\sqrt{26}$ _____ 4.982   A. <   B. >   C. =

47. Which of the statements is true for the pictured triangles?

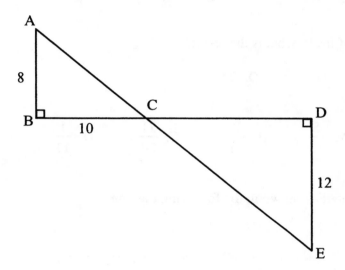

A. $CD = 18$ B. $\dfrac{AB}{AC} = \dfrac{12}{CD}$ C. $\dfrac{8}{12} = \dfrac{10}{CD}$ D. $\dfrac{DC}{BC} = \dfrac{AC}{AB}$

48. For each of the statements below, determine whether $x = 2$ is a solution.

i. $(x - 2)(3x + 1) = 4$

ii. $|1 - 5x| > 3$

iii. $3x^2 = 36$

A. i, ii, and ii   B. ii and 3 only   C. iii only   D. ii only

49. If $x = y^2 - 2z^2$ find x when $y = -2$ and $z = 3$.

A. -5 B. -14 C. 14 D. 22

50. Study the given information. In each figure, S represents the sum of the measures of the interior angles.

 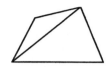 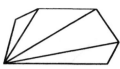

3 sides
1 triangle
S = 180°

4 sides
2 triangles
S = 360°

6 sides
4 triangles
S = 720°

Find S, the sum of the measures of the interior angles of a 15-sided convex polygon.

A. 2700°　　B. 1500°　　C. 2340°　　D. 3060°

51. When a number is increased by twice its square, the result is the number increased by 2. Which equation should be used to find the number, x?

A. $2x^2 + 2 = 2x$　　B. $x + 2x^2 = x + 2$　　C. $2x^2 = 2x$　　D. $(x + 2)^2 = x + 2$

52. Select the conclusion that will make the argument valid.

If I get another speeding ticket I will lose my license. I haven't lost my license.

A. I got another speeding ticket.

B. If I lose my license then I got another speeding ticket.

C. I didn't get another speeding ticket.

D. If I lose my wallet then I lose my video rental card.

53. Sixty percent of the students in a certain course are sophomores and 25% of the sophomores are taking the course as an elective. If one student is randomly selected, what is the probability that he or she is a sophomore who is taking the course as an elective?

A. .15　　B. .65　　C. .25　　D. .025

54. Study the figure showing the regular octagon. Then select the formula for computing the total area of the octagon.

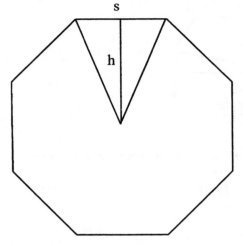

A. Area = 4sh   B. Area = 8sh   C. Area = 8(sh)^2 D. Area = 8(s + h)

55. Select the units of measure that would be appropriate for measuring the area of the circular surface forming the top of the cylinder shown below.

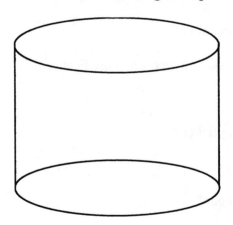

A. cubic centimeters   B. meters

C. inches   D. square feet

ANSWERS TO CLAST EXERCISES

1.1   1. C   2. D   3. A   4. B   5. D   6. C

1.2   1. C   2. A   3. A   4. A   5. A   6. B   7. D   8. C   9. C   10. B 11. C
     12. C

2.2,2.3 1. C   2. B   3. A   4. C   5. D   6. D

3.1   1. D   2. B   3. C   4. B   5. A   6. D   7. A   8. C   9. D   10. A 11. B
     12. C

3.5   1. C   2. B   3. B   4. C   5. C   6. D   7. A   8. C   9. B   10. D 11. A
     12. D 13. A 14. D 15. B 16. C

3.6   1. A   2. B   3. D   4. D   5. C   6. A   7. D   8. D   9. B   10. A 11. A
     12. C 13. D 14. A 15. A 16. C 17. C 18. C 19. B 20. C 21. D 22. C
     23. A 24. D

3.7   1. D   2. C   3. D   4. D   5. C   6. A   7. D   8. D   9. C   10. C

4.1   1. B   2. D   3. C   4. D   5. C   6. A   7. C   8. A   9. D

5.1   1. C   2. B   3. D   4. C   5. A   6. C   7. B   8. D   9. A   10. D

5.2   1. A   2. C   3. B   4. D   5. A   6. D   7. B   8. B   9. D   10. C

5.3   1. C   2. D   3. A   4. D   5. B   6. B   7. C   8. A   9. C   10. D 11. A
     12. C 13. B 14. C 15. A 16. B 17. D 18. B 19. D 20. B 21. C 22. A
     23. C 24. B 25. C 26. A 27. B 28. C 29. D 30. C 31. D 32. A

5.4   1. B   2. C   3. D   4. A   5. D   6. C   7. D   8. B   9. C   10. A 11. C
     12. B 13. A 14. D 15. B 16. D 17. C 18. B 19. A 20. B 21. B

5.6   1. B   2. A   3. D   4. B   5. C   6. C   7. D   8. A

5.7   1. B   2. D   3. C   4. D   5. C   6. A   7. B   8. B   9. D   10. B

6.1   1. B   2. D   3. C   4. A   5. D   6. D   7. A   8. B   9. B   10. C 11. A
     12. D 13. C 14. B 15. B 16. B 17. D 18. C 19. C 20. A

6.2   1. B   2. A   3. D   4. B   5. D   6. B   7. D   8. C   9. A   10. D 11. B
     12. C 13. A 14. C 15. D

6.3   1. C   2. D   3. B   4. C   5. B   6. C   7. A

KEY

6.4   1. B   2. D   3. C   4. A   5. B   6. C   7. D   8. C   9. D   10. C

6.5   1. B   2. A   3. D   4. C   5. A   6. D   7. A   8. C   9. B   10. B   11. C
12. A   13. C   14. D   15. A

6.6   1. C   2. A   3. C   4. B   5. D   6. B   7. C   8. D   9. D   10. C   11. C
12. D   13. C   14. A   15. C   16. D   17. B   18. B   19. D   20. A   21. A   22. A
23. C   24. B   25. B   26. C   27. A   28. D   29. C

7.1   1. B   2. A   3. D   4. B   5. C   6. B   7. A   8. C   9. D   10. C

7.3   1. A   2. C   3. D   4. B   5. A   6. B   7. C   8. C   9. B   10. D   11. C
12. B

7.4   1. D   2. B   3. A   4. D   5. D   6. C   7. C   8. D   9. A   10. C

8.1   1. D   2. B   3. B   4. C   5. C   6. A   7. C   8. C   9. A   10. D   11. A
12. A   13. B   14. A   15. B   16. C   17. D   18. B   19. B   20. D   21. C   22. B
23. A   24. B   25. B   26. D   27. A   28. A   29. C   30. D   31. B   32. C   33. D
34. A   35. D   36. B   37. D   38. C   39. A   40. D   41. B   42. C   43. B   44. C
45. C

9.1   1. A   2. C   3. A   4. B   5. D   6. B   7. C   8. D

9.2   1. B   2. C   3. D   4. A   5. D   6. B   7. A   8. D   9. C   10. D   11. C

10.1   1. A   2. C   3. D   4. A   5. C   6. B   7. D   8. C   9. B

10.2   1. D   2. C   3. B   4. C   5. B   6. C   7. D   8. C   9. C   10. C   11. C
12. D   13. B   14. A   15. C   16. A   17. B   18. C   19. A   20. D   21. D   22. C
23. A   24. D   25. D   26. C   27. D   28. B

10.3   1. B   2. C   3. B   4. C   5. B   6. A   7. D   8. C   9. D   10. A   11. B
10.4   12. D   13. A   14. C   15. D   16. B   17. B   18. C   19. B   20. A   21. C   22. B
23. D   24. C   25. A   26. C   27. D   28. C   29. C   30. D   31. B   32. C   33. C
34. B   35. A   36. B   37. D   38. A   39. C   40. B

10.4   1. B   2. A   3. C   4. B   5. C   6. C   7. A   8. C   9. A   10. B   11. B
10.5   12. D   13. A   14. D   15. B   16. D   17. C   18. B   19. D   20. C   21. B   22. D
23. A   24. C   25. B   26. C   27. C

11.1   1. B   2. A   3. C   4. A   5. D   6. B   7. D   8. A   9. D   10. A

11.2   1. C   2. B   3. A   4. B   5. A   6. B   7. C   8. C   9. D   10. C

11.3 1. A 2. D 3. B 4. B 5. C 6. B 7. C 8. D 9. B 10. C 11. B
12. A 13. D

11.4 1. C 2. B 3. D 4. C 5. A 6. C 7. B 8. C 9. D 10. A 11. B
12. A 13. C

11.5 1. C 2. A 3. B 4. A 5. C 6. D

11.6 1. D 2. B 3. D 4. D 5. C 6. D 7. C 8. C 9. B 10. A 11. B
12. D 13. D 14. A 15. D

11.7 1. A 2. B 3. D 4. C 5. A 6. B 7. A 8. C 9. B 10. C 11. C
12. D 13. B 14. A 15. D 16. C 17. C 18. B

12.1 1. B 2. D 3. C 4. B 5. A 6. C 7. C 8. D 9. B 10. C 11. B
12. A 13. B 14. B 15. D

12.2 1. A 2. C 3. B 4. C 5. A 6. D 7. B 8. D 9. D 10. C 11. D
12. B 13. A 14. A 15. D 16. C 17. C 18. D 19. D 20. C 21. C 22. C
23. C 24. A 25. D 26. A 27. D

12.4 1. B 2. A 3. A 4. D 5. C 6. A

12.5 1. C 2. B 3. D 4. A 5. A 6. C

Practice CLAST A
  1. A 2. C 3. D 4. C 5. A 6. B 7. B 8. C 9. B 10. D 11. B
12. C 13. C 14. D 15. B 16. D 17. B 18. C 19. A 20. D 21. D 22. C
23. A 24. D 25. A 26. C 27. D 28. C 29. A 30. D 31. C 32. A 33. D
34. B 35. C 36. A 37. A 38. C 39. C 40. A 41. B 42. D 43. A 44. B
45. B 46. B 47. C 48. C 49. D 50. D 51. C 52. B 53. D 54. D 55. A

Practice CLAST B
  1. D 2. A 3. A 4. D 5. B 6. C 7. D 8. A 9. C 10. D 11. B
12. C 13. A 14. D 15. C 16. D 17. D 18. C 19. B 20. A 21. D 22. A
23. C 24. B 25. C 26. B 27. C 28. C 29. A 30. C 31. C 32. A 33. B
34. C 35. C 36. B 37. B 38. C 39. A 40. C 41. B 42. A 43. C 44. C
45. B 46. B 47. C 48. D 49. B 50. C 51. B 52. C 53. A 54. A 55. D